THE ART IMPERATIVE

THE ART IMPERATIVE

THE SECRET POWER OF ART

PHILLIP ROMERO, MD

To order additional copies of this book, contact:
Xlibris Corporation
1-888-795-4274
www.Xlibris.com
Orders@Xlibris.com
78163

CONTENTS

Preface

Like most artists, I began making art as a child—I had been dubbed the "class artist" by the third grade and knew that art would be a part of my life forever. I struggled as a young man as to where I would get special training—art school or medical school. When I was accepted to med school, I could not say no, but I continued to make art all the while—exhibiting and selling my first mature artworks during my years at University of Texas Medical Branch in Galveston, Texas. I befriended my first "professional artists" while there: Michael Tracy, Thomas Downing, Joe Glasco, and the young art student, Julian Schnabel.

During medical school, my interest in cross-cultural study took me to India for three months to study at the Yoga Institute in Bombay and Dharamsala to study Tibetan medicine, art, meditation, and Buddhism. I made another medical anthropology excursion to Pan Am University to study curanderismo (indigenous folk medicine) in South Texas and North Mexico.

Medical training, my study of cross-cultural medicine, and my newfound interest in Buddhist philosophy and meditation inspired my art—anatomy lab, histology, radiology, Tantric mandalas, Tibetan sandpainting, etc. As a native of Port Arthur, Texas, home of Robert Rauschenberg and Janis Joplin, I felt free to experiment with art-making, taking things from medicine and incorporating them into work—X-rays, my own EEG tracing, plastic tubing, surgical gauze, plaster casting, syringes, neon lighting, a found dog skull, etc.

In 1982 my daughter was born in New York City—I was a fellow in child psychiatry at that time—and I was deeply curious about how we human beings developed our passion for art. Watching her birth inspired an awakening to my own "womb-envy"—"I make art because I am envious of my wife's ability to create a child"—the ultimate artwork.

I took that experience and imagined how prehistoric man may have felt at the mystery of childbirth. I set about writing a developmental psychology essay entitled "Men, Their Painting, and Their Fear and Envy of Women"—focusing on the cave paintings of Lascaux, I speculated that the origins of men's art-making were inspired by the mystery of creating life and by his early gender-differentiation from the mother in the 2-4-year-old phase. Little boys, I speculated, had a much deeper drive than girls to create some kind of product—painting, drawing, building, inventing, etc—in order to prove their value. Boys' creative activity had greater survival value to hunter-gatherers than girls, who were endowed with the mother's built-in capacity to have children, their miraculous creative capacity giving them special status in the community. If women's role was to bear and nurture children, men's role would be to nurture and protect the mother-child unit. Although it was never published, my professors encouraged me to pursue it with interviews with adult artists.

From my essay emerged *The Art Imperative*—a book that aimed to explore the early development of male and female artists' personal motivation to become an artist. By 1986 I was negotiating with Abrams to publish the book, and I had interviewed Audrey Flack, Louis Bourgeois, Michael Tracy, Gary Stephan, Joe Glasco, Tom Downing, Jeffrey Lew, and the interview with Andy Warhol had begun. Andy and I were planning to finish the interview when he was hospitalized for gall bladder surgery—he died in that tragic hospitalization—I was unable to complete the project at that time and put it away for another time in the future.

In 1998 E. O. Wilson published *Consilience: The Unity of Knowledge*, and he was keynote speaker at a three-day conference with the same title sponsored by the New York Academy of Science. It was clear from this groundbreaking conference that we were entering a new period of bridging domains of knowledge. My clinical interest in brain development, creativity, and resilience to adversity were amplified by the work of Bruce McEwen of Rockefeller University. His presentation and book, *The Hostage Brain,* triggered my return to *The Art Imperative,* this time with fifteen years of clinical experience and huge leaps in understanding brain development as a result of fMRI (functional brain imaging technology) studies of the functioning brain. In correspondence with E. O. Wilson, I proposed the concept of Consilient Art as an agent of cultural resilience, to which he replied a validating "sounds true."

In the last five years, conversations about Consilient Art with Joe LaPlaca, artist-journalist-art producer and cofounder of All Visual Arts, London, have taken the core ideas in this little book, twenty-five years in the making, and inspired *The Art Imperative Project*. In this collaboration, we are exploring the role of Consilient Art currently being created by such artists as Paul Freyer, Damien Hirst, Matthew Richie, and many others, as agents for cultural resilience to contemporary adversities in cross-cultural conflict, art business, and globalization of the world economy. Multimedia projects are in the planning stage at this writing.

Phillip Romero, MD, New York, January 2010

Introduction: The Secret of the Art Imperative

Art presents a paradox: Although it seems to have no apparent utilitarian function, its ubiquitous existence throughout the world suggests the opposite—there must be some secret power encrypted in art that promotes survival value—an Art Imperative.

Art-making emerged in human history over thirty thousand years ago, evidenced by the wondrous cave paintings of Lascaux, Altamira, Chauvet, and others (fig. 1). Today, art is a universal phenomena in all cultures that distinguishes *Homo sapiens* from all other creatures. This book proposes that art-making plays a critical role in human survival. To support this proposal, consilient (unity of knowledge) facts and fact-based theories are integrated from evolutionary sociobiology, developmental neuroscience, attachment theory, consciousness science, and complex systems theory—central to the proposition are the theories of Buddha, Charles Darwin, John Bowlby, and the science and theories of many others.

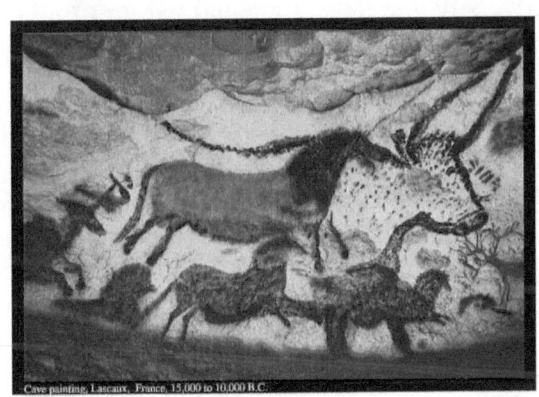

Cave painting, Lascaux, France, 15,000 to 10,000 B.C.

Fig 1 Lascaux

For individuals, art-making facilitates the practice of the human brain's capacity for creative problem solving from early development across the life cycle. For cultural systems from the family to society, art facilitates social connectedness and the creation of collective meaning across the generations, the essentials for adaptation to adversities that challenge the integrated function of all cultural systems. Human beings create abstract stories of their own meaning—their collective life mission beyond mere survival—that empower enormous capacity to face overwhelming adversities from nature or other human beings. Meaningful social connectedness equals survival—from mother-infant attachment to the integrity of global cultures.

The Art Imperative Hypothesis

A hypothesis is proposed that the creating of abstract representations of complex meaning in art provides individuals and cultures with the opportunity for private reflection and social discourse—the critical cognitive, emotional, and social functions that empower human beings individually and collectively with the capacity to adapt to the inevitable stresses that can disconnect individuals and disorganize cultures. The theories of Buddha, Charles Darwin, and John Bowlby offer a new map of territories that include the science of consciousness, the plasticity of brain function, the evolution of social connectedness for resilience to adversity, the developmental nature of attachment and loss, and the critical role of creativity and compassion in the survival and diversity of human cultures. Bridging these domains of knowledge offers a consilient foundation for a new understanding of the human urge to make art. Recent advances in the neuroscience of brain plasticity for creative problem solving advance ideas of how art can be reframed as a human imperative critical to individual adaptation, cultural resilience, and species survival.

Buddha, Darwin, and Bowlby: A Consilient Approach to Art

BUDDHA'S THEORY OF EMPTINESS

Two thousand five hundred years ago, Buddha taught the Four Noble Truths that diagnosed the essence of human suffering arising from our attachment to life and our fear of loss of these attachments. He prescribed the Eightfold Path with three components: wisdom, ethics, and mental discipline or meditation to contemplate these truths. Training the mind in effortful reflection is critical to the prescription that liberates one from the suffering of life. Insight and compassion meditation offer a deep experiential understanding of *Shunyata*—usually translated as "Void" or "Emptiness," but sometimes more positively as "Openness." The philosophical nuances of this term are complex and paradoxical. Emptiness does not mean nihilism (the denial that anything exists) or describe some absolute entity that underlies appearance. Rather, things are "empty" in the sense of lacking independent, permanent existence. Ultimately, Shunyata is a phenomenological term for the experience of pure consciousness. This experience derives from the attainment of the capacity for selfless awareness and total freedom from the distortion and limitations of the ego or sense of self and its conceptual understanding— it is "empty" or "void" of all such particular characteristics. The Heart Sutra contains Buddhism's most celebrated paradox:

> form is emptiness and the very emptiness is form; emptiness does not differ from form, form does not differ from emptiness; whatever is form, that is emptiness, whatever is emptiness, that is form, the same is true of feelings, perceptions, impulses and consciousness . . . . it is because of his non—attainment

that a Bodhisattva, through having relied on the Perfection of Wisdom, dwells without thought-coverings. In the absence of thought—coverings he has not been made to tremble, he has overcome what can upset, and in the end he attains to Nirvana . . .

Buddhist philosophy derives from both analytic reasoning and the experience of this paradox during meditation. Phenomenon and experience are seen to emerge in consciousness as an ever-changing dynamic of cause and effect (karma) with relative and impermanent meaning. Nothing has independent meaning in itself. This concept reflects some of the latest theories in physics, complex systems theory, and cognitive neuroscience. Buddhist brain-training techniques that cultivate mindfulness and a compassionate attitude are being explored by neuroscience and appear to have a major impact on health and well-being. Recent brain imaging studies have discovered that the nonfocused mental activity of "daydreaming" (mind-watching) actually enhances activity in the problem-solving prefrontal cortex, a critical part of the neural networks for creativity.

Perhaps the greatest contribution of Buddhist mindfulness practice is the attainment of the deep acceptance of the fact of impermanence of all things—attachment is natural, separation and loss are painful, and when one awakens to the miraculous beauty of being human, these experiences can be understood as the timeless beauty of being a conscious, sentient being.

DARWIN'S THEORY OF EVOLUTION

Charles Darwin's controversial theory of evolution advanced scientific thinking and practice more than any other. His theory proposed three core ideas:

- A species is a population of organisms that interbreeds and has fertile offspring.
- Living organisms have descended with modifications from species that lived before them.
- Natural selection explains how this evolution has happened:
  a. More organisms are produced than can survive because of limited resources.

b. Organisms struggle for the necessities of life; there is competition for resources.

c. Individuals within a population vary in their traits; some of these traits are heritable—passed on to offspring.

d. Some variants are better adapted to survive and reproduce under local conditions than others.

e. Better-adapted individuals (the "fit enough") are more likely to survive and reproduce, thereby passing on copies of their genes to the next generation.

f. Species whose individuals are best adapted survive; others become extinct.

IMPROVISATION AND COLLABORATION: THE EVOLUTIONARY SECRET OF ART

Evolution theory focuses on the capacity for adaptation to adversity as the optimal attribute for any species to survive and for humans to pursue well-being. All human beings struggle with stress across the life cycle, some much greater than others. Adverse situations trigger the body's allostasis system (the organisms' built-in capacity to change itself for adaptation), commonly known as the stress response. This episodic stress reaction sparks resilience—increased heart rate, blood flow to the muscles, increased immune response—but if the stress system is chronically triggered, the neurochemical effects are toxic.

Darwin proposed that the capacity for resilience to adversity of any species emerges from the ability to adapt to the threatening situation. Optimal adaptation is accomplished by "improvisation and collaboration." These two critical ingredients for survival form the essence of the secret power of art—by facilitating improvisation and collaboration in culture, art empowers cultural resilience to the stresses that perpetually threaten the cohesion of society. Artists' creative imagination improvises new forms and images, inviting reflection and discourse—a virtual brain-gymnasium for the imagination of the viewer.

Just by looking at new art, the viewer encounters charged images that startle, engage, repulse, seduce, mystify, and inspire wonder and curiosity—very little effort is required beyond gazing at the art object. The success of art emerges through the collaboration of collectors, craftspeople, museums, critics, and the public. Art-making mobilizes many people in different walks of life within a culture, brings them together through the exhibition of the art object, and invites all segments of culture to participate in the private and public reflection on art. Art is the product of individual improvisation and collective collaboration—the potential to connect diverse elements of society with abstract imagery implicitly reflects the dynamic aspects of individuals and their culture.

BOWLBY'S THEORY OF ATTACHMENT

In the 1960s, John Bowlby shook up the psychoanalytic world by challenging Freud's theories with his attachment theory.

Sigmund Freud, a great believer in Darwin, placed the sexual urge at the center of species survival. By focusing on the role of secure attachments rather than the sexual drive as the foundation for optimal human adaptation, Bowlby opened the way for scientific study of cognitive, emotional, and motivational development.

Bowlby's interest in the ethology theories of Konrad Lorenz concerning organization and elicitation of individual and social behavior patterns in animals linked Darwin's and Freud's theories in a new framework. His focus on the natural processes of attachment, separation, and loss probed the mechanisms of successful and unsuccessful adaptation of the child's development into adulthood. The scaffold of challenges that face secure attachment across the life cycle triggers creative responses in the course of all human relationships. It is in these creative responses to the ever-changing adversities of life that individuals, cultures, and species survive.

The surrealists were enchanted by Freudian notions of the unconscious and the automatic processes of creativity. Like their counterparts in psychoanalysis, artists explored their own unconscious imagery, hoping to leave a trace of universal processes that connect human beings across time and culture. The artistic exploration of the inner self navigated two world wars, crossing the Atlantic Ocean, and emerged in America in the abstract expressionist movement that focused more exclusively on the painting of emotion and the existential confrontation with chaos. Like their creative counterparts in the sciences who explore the nature of the material world, artists explore the nature of being human, the perceptions, emotions, relationships, joys, and adversities that challenge each and every one of us.

A CONSILIENT APPROACH TO ART

In 1998 evolutionary biologist E. O. Wilson published his Pulitzer-winning *Consilience: The Unity of Knowledge,* in which he discusses methods that have been used to unite the sciences and might in the future unite them with the humanities. Wilson's call for consilience reaches out to all disciplines of knowledge—consilience being defined as "Literally a jumping together of knowledge by the linking of facts and fact-based theory across disciplines to create a common groundwork for explanation."

Buddha devoted himself to understanding the nature of human suffering as a product of mind and teaching a way toward the end of suffering. Darwin spent a lifetime probing the evolution of "endless forms most beautiful" that led to the human race and explicating the mechanisms by which such an evolution could take place. Bowlby's attachment theories made possible the scientific study of emotion, social relationships, and culture—social neuroscience—and helped identify the mechanisms of social attachment that give rise to our sense of self and security in an uncertain world. Taken together, these theories provide a new foundation for explaining the mysterious nature of the human urge to create art—the Art Imperative.

Human development, survival, and the capacity for resilience to ever-changing environmental stressors rely on our ongoing creativity. From cradle to grave, from one generation to the next, from the rise and fall of cultures, human beings face new and daunting challenges to their survival and well-being. With each new concrete challenge come new creative solutions. What does not change across the millennia is the human need to make sense of its existence, to create meaning that sustains and fulfills our insatiable lust for life. We are driven to solve the riddle of being human by creating meaning in each new generation. For thirty thousand years, human beings have been making art in an ongoing process critical to the creation of our meaningful existence.

PART I

ART AND STRESS

One can have no smaller or greater mastery than mastery of oneself.

—Leonardo da Vinci

Chapter 1.0

THE ART IMPERATIVE:
CREATIVITY AND RESILIENCE

Only the true self can be creative and only the true self can feel real.

—D. W. Winnicott

1.1 BIRTH OF THE SELF: ART-MAKING AND MAKING ART

"Art-making" is used here to describe the spontaneous activity for self-expression in the developing child—a reciprocal process with mother and caregivers that facilitates secure attachments. "Making art" refers to the more mature brain's wish-driven design to explore, experiment, and practice art-making—thus reconnecting with the authentic and spontaneous experience of early childhood. This chapter explores how the process of art-making facilitates self-regulation and how the making of art connects artists with their authentic origins of expression from early childhood and communicates complex images to society for reflection and contemplation. Artists and art viewers are connected by art in a reciprocal social relationship—one function of this connection being to secure attachment and resilience to adversity throughout the life cycle of Individuals.

Integrating knowledge about evolutionary resilience, brain development, and creative adaptation to adversity opens a new, consilient view of human art-making. Like the evolution of the capacity for language, human beings evolved the capacity to create art as a critical tool for individual and cultural survival—an Art Imperative. By translating Buddha's Four Noble Truths into the language of evolution and social neuroscience, a consilient framework for exploring this Art Imperative can be constructed.

Buddha's Four Noble Truths	Evolution and Social Neuroscience
1. Life is suffering.	Life is stressful.
2. The cause for suffering is attachment and the fear of loss.	The cause of stress is rooted in adverse experiences that produce insecure attachments to people, places, and things.
3. There is a solution to suffering, which is to embrace the "Emptiness" of all things and the truth of impermanence.	Accepting the truth of causal relativity and impermanence with gratitude for the social connections that kindle meaningful life experiences is protective against stress and imbues resilience, creativity, joy, and longevity.
4. There is a method to apply the solution, which is the Eightfold Path.	A consilient approach to art and art-making offers new methods for adaptive resilience to adversity for individuals, cultures, and global systems.

A scientific view emerges: Life is stressful. There is a cause for stress. There is a solution. Creating art and experiencing art play a key role for the individual, culture, and the human species in navigating the stresses of life. Stress can derail child development and disconnect individuals from themselves, from each other, and from society. Art links individuals with culture in a reciprocal relationship that facilitates the adaptive resilience of both during times of adversity. Unlike governments or religions, art and art-making are not intrinsically dogmatic belief systems, although art can be employed by dogmas for propaganda. This freedom from preconceived ideas of what art should be or should not be creates a unique place for art in individual psychology and cultural evolution. The private experience of art-making and art viewing connects artist with individuals in society in nonconscious ways that can change the viewer through a variety of cognitive and emotional experiences.

AUTOPOIESIS: MATURANA'S THEORY OF SELF-MAKING

Neuroscientist-philosopher Humberto Maturana proposed that "Living systems are cognitive systems, and living as a process is a process of cognition. This statement is valid for all organisms, with or without a nervous system . . . living beings are characterized in that, literally, they are continually self-producing. We indicate this process when we call the organization that defines them an *autopoietic organization*."

From single-cell organisms to the complex human beings to social organizations, all living systems involve ongoing creative relationships between their organization and structure. Human beings remake themselves in a process far beyond sexual reproduction, which produces the structure of an infant, but the organization of the infant into a human being requires a complex social system of parenting, family, community, and culture. These relationships organize the structure of the body into a functional language-culture-making person. The story for the body that narrates it into a person begins in the mother-infant relationship. Maturana uses the term *languaging* to identify the complex cognitive-emotional-social process that organizes the relationships between people.

1.2 WINNICOTT'S THEORY: CREATIVITY AND THE SELF—A RECIPROCAL RELATIONSHIP

All children make art. Art-making is one of many processes in childhood play, and art is the product of this activity. Like language, art-making does not spring from the child's brain *de novo*; rather, it emerges from creative reciprocal play with parents and others. Immediately after its birth from the oceanic womb, the brain must begin adapting to its new terrestrial environment. John Bowlby theorized that a secure attachment is the primary adaptive drive of the mother-infant relationship.

Bowlby's colleague, pediatrician-psychoanalyst D. W. Winnicott, added a critical understanding to this process with the phrase, "There is no such thing as a baby." Winnicott focused on the interdependent system of the mother-infant relationship as cocreating the emerging sense of self in the baby. The art of parenting is no perfect practice, and under optimal circumstances, this process is facilitated by what Winnicott calls "good-enough mothering," in which the mother creates a secure "holding environment" for the infant to explore and discover the external world and develop a secure sense of identity.

The middle ground between "objective reality" (also called the "not-me") and "subjective omnipotence" (the "me") is what Winnicott calls the "transitional experience." *Transition* refers to the process that aids the child while the mother separates, and the experience is a transitional zone between the self and the real world.

Central in the transitional experience is the "transitional object" that inhabits this zone, which to the infant represents the mother or her breast when she is absent. This object can alternatively be referred to as the first "not-me" possession—a teddy bear, a blanket, etc. Winnicott theorizes that the child does not experience this object as created by themselves, nor as entirely detached, but instead the transitional object is a fantasy. Created by the brain's efforts to integrate the emotional and cognitive perceptions of separation and the biological-instinctive fear of annihilation, the transitional object functions as a bridge between subjective omnipotence of "me" and objective reality of "not me"; it is a compromise to make sense of the difference between feeling secure and the awareness of the insecurity of separation-annihilation.

The transitional object is charged with a double meaning. It is a way for the child to maintain a connection to the mother while it learns that the stress of separation from mother as she progressively distances herself can be tolerated. The transitional object helps the child's brain self-regulate its anxiety while finding a balance between his or her own subjectivity and accommodation to others. The transitional experience allows the child to practice self-soothing behaviors to adapt to the objective reality that the mother is a separate being. It is from this universal developmental process that the brain begins its self-making journey and the art-making potential emerges.

1.3 CREATIVITY AND ART: TRANSITIONAL SPACE AND TRANSITIONAL OBJECTS

Compared with nonhuman primates, human beings have a very prolonged period of dependency on caregivers before gaining autonomy. The payoff for this process is the development of a brain designed for abstract, creative computations that can coordinate complex social behaviors and dominate all other species. The capacity to recall the past, imagine the future, and design a more-effective strategy for survival in the present facilitates the creation of culture and the possibility for the pursuit of well-being.

The brain's creativity begins in the mother-infant relationship with the adaptive response to the stress of separation-individuation. Key to the navigation of this process is what Winnicott calls the transitional experience. This drama of interactions with the transitional object unfolds in the transitional space—the abstract zone where the developing brain utilizes its embryonic imagination to help the child self-soothe his separation anxiety and instinctual fear of annihilation in the absence of parenting.

This capacity for self-soothing is critical for adaptive resilience to adversity across the life cycle. Artists access the transitional experience in the making of art to stay connected with their own primordial origins of self and charge their art-works with authenticity. An example of this is found in Barnett Newman's description of the Native American Kwakiutl artist:

> The Kwakiutl artist painting on a hide did not concern himself with the inconsequentials that made up the opulent social rivalries of the Northwest Indian scene; nor did he, in the name of higher purity, renounce the material world for the meaningless materialism of design. The abstract shape he used, his entire plastic language, was directed by a ritualistic will toward metaphysical understanding. The everyday realities he left to the toy makers; the pleasant play of non-objective pattern, to the women basket weavers. To him an abstract shape was a living thing, a vehicle for an abstract thought-complex, a carrier of the awesome feelings he felt before the terror of the unknowable. The abstract shape was, therefore, real rather than a formal "abstraction" of a visual fact, with an overtone of an already—known nature. Nor was it a purist illusion with its overload of pseudoscientific truths.

Newman's interpretation of the Kwakiutl artist's motivation and experience in making paintings sounds almost exactly like Winnicott's description of the transitional experience where the painting is the transitional object. By making an abstract image of his encounter with the "terror of the unknowable," the artist's art-making becomes a transitional experience that functions to soothe the artist's existential anxiety. By recording his experience in a painting, he can demonstrate his journey to the edge of security for his community. His artwork is an artifact of his emotional and cognitive journey, a visualization of his experiential truth of surviving his encounter with the "terror" of aloneness or meaninglessness (the equivalent of the separation-annihilation anxiety of childhood) (fig. 2). When the artwork is accepted by the community, it becomes charged with sacred meaning to ward off the evil of separation or annihilation for the entire community, providing it with a sense of security. What was personally experienced by the artist becomes a cultural artifact with sacred meaning for the community.

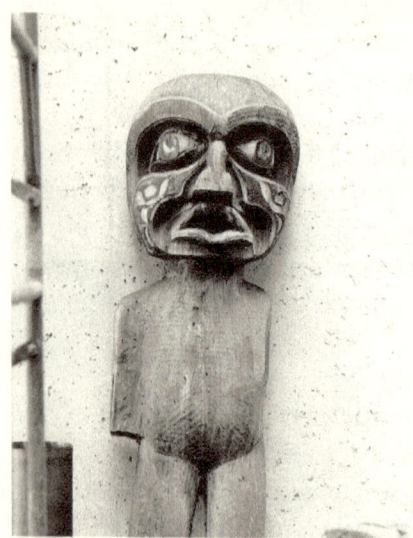

Fig 2 Kwakiutl Figure

1.4 ART AND THE TRUE SELF: THE SEARCH FOR AUTHENTICITY

Winnicott asserted, "Only the true self can be creative and only the true self can feel real." For Winnicott, the "True self" is instinctive—the biological core of the personality. Rooted in the infant's capacity for spontaneous self-expression, a True self has a sense of integrity, of connected wholeness. As the infant brain gains control over its bodily functions and begins to mobilize, a spontaneous self emerges with a mission—explore, experiment, practice—to gain a sense of safety and mastery in its environment. The experience of spontaneity and aliveness is the heart of authenticity and the seed of a potential True self. Its growth is dependent on the nurturing attunement of the mother.

Winnicott described these behaviors and attitudes as the good-enough mother who is repeatedly responsive to the infant's "illusion of omnipotence" and to some extent makes sense of it. The True self flourishes only in response to the repeated success of the mother's optimal responsiveness to the infant's spontaneous expressions.

Like the True self, the emergence of the False self results from the mother-infant reciprocal relationship when the "not good-enough mother" is too self-preoccupied and does not sense and respond optimally to her infant's needs. Instead, she substitutes her own self-serving gestures with which the infant complies. The infant's repeated compliance becomes the ground for the earliest mode of the False self's existence.

Through this False Self, the infant builds up a false set of relationships and, by learning to imitate caregivers, can even attain a show of being real so that the child may grow up to be just like the mother, nurse, aunt, brother, or whoever at the time dominates the scene. The primary function of the False self is adaptive, to protect the True self from the threat of abandonment by the self-centered "not good enough mother." This is an unconscious process: the False self comes to be mistaken for the True self to others, and even to the self. Winnicott described the False self in adulthood as when the person has to comply with external rules, such as being polite or following social codes, then a False self is used. The False self is a mask of the false persona that constantly seeks to anticipate demands of others in order to maintain the relationship—the False self is a people pleaser to the extreme of self-neglect. Even with the appearance of success and of social gains, there will be feeling of unreality, a sense of not really being alive, and that happiness doesn't or can't really exist due to the disconnection with the True self.

1.5 BARNETT NEWMAN: AUTHENTIC ART AND CREATIVE RESILIENCE

The creation of art is rooted in the development of the sense of self. A critical aspect of timeless art is reflected in its originality and authenticity—the trace of the artist's True self. Like DNA, authentic art is identifiable by characteristics unique to the artist. From the Paleolithic cave art to the abstract expressionist artist Barnett Newman's *Onement I* (fig. 3), authentic art can never be duplicated—it can be copied, but the authenticity of True art springs from the True self, and its originality is found in the overt and subtle idiosyncrasies of the artist. A consilient approach to the early artworks and writings of Barnett Newman reveals a reciprocal linkage between the artist's private emotional truth and his cultural-historical context. Newman's personal identity development as an artist and his creative struggle with his historical context demonstrate the specialized place that artists can occupy in culture.

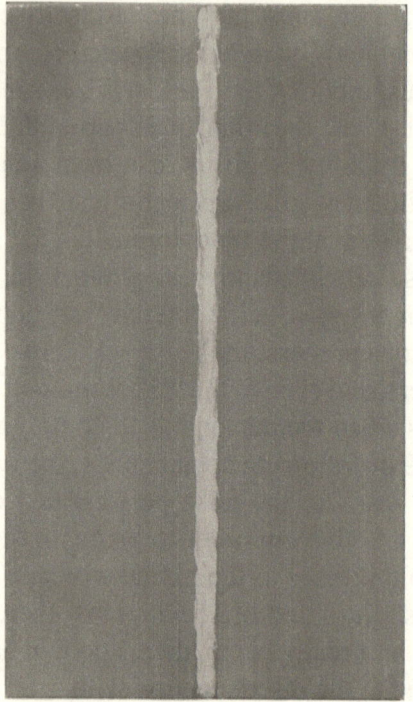

Fig 3 Barnett Newman *Onement I*

Newman's seminal painting *Onement I,* his first mature "zip" painting, marked what would be his defining style and contribution to art history. Having destroyed all of his art before 1944, he said,

> I recall my first painting—that is, where I felt that I had moved into an area for myself that was completely me—I painted on my birthday (January 29) in 1948.

Born in 1905, Newman began making art seriously after study at the Art Students League in New York City in 1922 at age seventeen. In response to World War II, he stopped making art in 1940 reflecting,

. . . we felt the moral crisis of a world in shambles, a world devastated by a great depression and a fierce world war, and it was impossible at that time to paint the kind of paintings we were doing—flowers, reclining nudes, and people playing the cello. At the same time we could not move into the situation of a pure world of unorganized shapes and forms, or color relations, a world of sensations. And I would say that for some of us, this was our moral crisis in relations to what to paint. So that we actually began, so to speak, from scratch, as if painting were not only dead but had never existed.

Newman's moral dilemma about his world has a powerful influence on the question, "What to paint?" As the artist enters the abstract world of image making, of facing a blank canvas and creating a world out of his imagination, he brings with him, whether consciously or nonconsciously, the reality and stress of his everyday life, whether petty or catastrophic. Newman wrote in 1944, "On Modern Art: Inquiry and Confirmation,"

. . . modernism brought the artist back to first principles. It taught that art is an expression of thought, of important truths, not of a sentimental and artificial "beauty". It established the artist as a creator and a searcher rather than a copyist or maker of candy.

Newman's personal origin as an authentic artist takes place during his early 40s (1944-48). With an educational background in philosophy and biology, Newman, was on a personal and professional quest for his own truth as a painter. His world was in chaos, divided and traumatized by the realization that mankind had created a nuclear weapon capable of mass destruction that was more terrifying than almost anything in nature. Science and the idealized quest for enlightenment had led to the dark side of creativity—the power to destroy everything. The division between creation and destruction had never been so globally evident in humankind's collective consciousness—human beings are capable of totally destroying themselves. In 1945 Newman wrote a private essay, "The Plasmic Image," with the opening statement,

The subject matter of creation is chaos . . . The present painter can be said to work with chaos not only in the sense that he is handling the chaos of a blank picture plane but also that he is handling the chaos of form. In trying to go beyond the chaos of the visible and the known world he is working with form unknown even to him.

Newman saw himself as an agent of his culture, in search of new ways to creatively adapt, to make sense of the chaos of his time, and as an artist, to leave a trace of his quest for truth and understanding. He theorized that "primitive" artists were motivated by the fear of the unknown, destructive forces of nature in the motivation for making art, and that the modern artist faced a similar kind of cultural reality—the man-made destructive force of the nuclear bomb. As one of the theoretical creators of the abstract expressionist art movement—which included Clyfford Still, Jackson Pollock, Ad Reinhardt, Mark Rothko, and others—he wrote,

> We know the terror to expect. Hiroshima showed it to us. We are no longer, then, in the face of a mystery. After all, wasn't it an American boy that did it? The terror has indeed become as real as life. What we have now is a tragic rather than terrifying situation . . . Our tragedy is a tragedy of action in the chaos that is society . . . and no matter how heroic, or innocent, or moral our individual lives may be, this new fate hangs over us.

Newman's writing provides a narrative of his personal views of his time and the moral position that faced the abstract expressionists. His experiments and explorations in drawing and painting in his gestational years demonstrate the process of a self in search of a form that could represent his personal truth—his True self—and connect it meaningfully with his culture. He called the stripes in his paintings "zips," declaring that

> I feel that my zip does not divide the painting. I feel it does the exact opposite. It does not cut the format in half . . . it unites the thing. It creates a totality . . . . [1]

---

[1] J. Strick, *The Sublime Is Now: The Early Work of Barnett Newman* (New York: Pace Wildenstein, 1994), 8-26.

For Newman, both the process and product of his great works involve the unification, the mending, and reintegration of self and culture. Although Jackson Pollock's action paintings (fig. 4) differ radically from Newman's subdued flat surfaces, he echoes Newman's sentiments that reflect a radical new approach to making art desperately needed by American and world culture to adapt to the nuclear world that emerged in WWII:

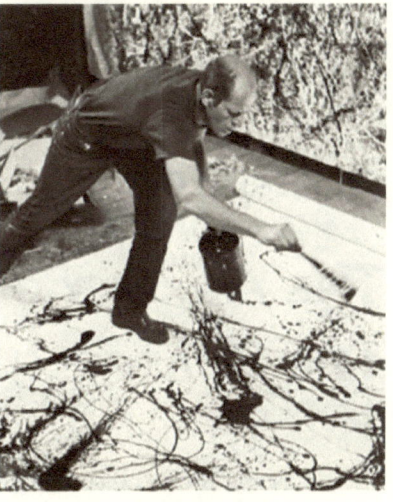

> The modern artist is working with space and time, and expressing his feelings rather than illustrating . . . . The modern artist . . . is working and expressing an inner world—in other words—expressing the energy, the motion, and other inner forces.

Fig 4 Jackson Pollock
Action Painting

Chapter 2.0

THE ANXIETY OF IMPERMANENCE: THE SCAFFOLD OF INSECURITY

> From the very core of our being, we desire contentment. Cultivating a close, warmhearted feeling for others automatically puts the mind at ease. It helps remove whatever fears or insecurities we may have and gives us the strength to cope with any obstacles we encounter.
>
> **—Dalai Lama**

Critical developmental hurdles challenge each human brain on its way to becoming a self-regulating, well-connected adult: attachment to the mother, socialization with peers, and individual identity development. Each of these steps stresses our brain-body efforts to form strong social bonds as nature pushes us toward becoming our selves. We experience this stress as anxiety: (1) separation anxiety during the formation of a secure infant-mother attachment, (2) social anxiety during the early childhood learning of shame, (3) identity and gender anxiety from toddlerhood to teen, (4) existential anxiety during the adolescent development of the capacity for abstract thinking. These four anxieties scaffold into humankind's number one insecurity and lifelong emotional enemy, what I call the anxiety of impermanence.

2.1 Art, Allostasis, and Affect Regulation

The allostatic system (stress-response system) connects the brain with the body and triggers the survival behaviors to freeze, fight, and flee from danger.[2] It is also intertwined with the urges to forage for food and find a sexual partner. The allostatic system is engaged when we are confronted with life-threatening events or minor surprises that startle us. It can be triggered by a sexy fantasy or a scary dream. The five Fs—freeze, fight, flight, feed, fornicate—are hardwired into our body and brain, and developmentally malleable with experience. They are regulated by the limbic structures in the midbrain. The critical allostatic structures in the brain are the amygdala, the hypothalamus, the pituitary, and the hippocampus. Key body structures include the autonomic nervous system, the adrenals and other endocrine glands, and the immune system. This allostatic system is interconnected with virtually every other part of the brain, from the lower "reptilian brain" that governs heart rate and respiration to the higher cortical levels, especially the prefrontal cortex (PFC) that are involved in decision making, abstract thinking, and reflective consciousness.

The allostatic system can also be modulated by our brain's powers of imagination, language, and art. We can actually train our brains to become more efficient in utilizing our stress system for more than survival and procreation. We have evolved the ability to override our knee-jerk survival impulses with creativity, reflection, and abstract narrative meaning (a story with a cause). We can learn to regulate our affective responses to stress by creating new, abstract stories that will strategically guide the body to a higher order of personal decision making and social cooperation for well-being and for humanity. When the PFC creates a narrative of personal or collective meaning, it can override the impulse to freeze-fight-flee—face the fear—and create new actions that form the basis of culture. Using language, conversation and compromise offer alternatives to fighting. The arts of music, dance, and art provide expressive forms to modulate fear, confusion, and anger.

Art is one of the most ancient and powerful attributes of humankind that facilitates ongoing adaptation to stressful events, empowers affect regulation, and kindles cultural resilience. The brain's efforts to navigate

---

[2]   B. McEwen, *The End of Stress as We Know It* (Washington, D. C.: Joseph Henry Press, 2004).

the stress of life begin in the creative play between infant and mother. Art emerges from the real and imagined actions, feelings, and thoughts as a means to practice adaptive resilience to adversity.

2.2 SEPARATION ANXIETY: SECURE OR INSECURE ATTACHMENT?

The earliest connection we make after birth is with the mother, our primary physical, emotional, and social attachment. Our body is traumatically pushed out of the womb—its secure, oceanic sanctuary— and born into a world of overwhelming sensation. Birth trauma is our first separation. Infants are totally dependent on the "holding environment" created by the attention and emotional attunement of "good-enough-mother"[3] to her baby. As infants, our primary anxiety, *separation anxiety,* springs from the ordinary disconnections that occur in our secure attachment to the mother.

A secure base of attachment[4] to mother and caregivers emerges from countless infant-mother experiences where the secure emotional bond and shared attentional connection is stressed and disconnected. The "good-enough mother" lays a foundation for stress regulation by attuned reconnection with her baby that actually alters DNA expression, which helps regulate the allostatic system over the entire life of the individual.

The repetitious practice and experience of the "connection-disconnection-reconnection" cycle provides the brain with the practice needed to shape a resilient stress-regulation system. Animal models demonstrate that a securely attuned mother-infant relationship tempers the allostatic system for life.[5]

The brain and body of the infant begin life in a relationship with the mother that has biological, social-emotional, and abstract narrative systems that are constantly interacting. This complex reciprocal system has been described with a "Mutual Regulation Model" that sees the infant as part of a dyadic communicative system in which the infant

---

[3]   D. W. Winnicott, *The Maturational Processes and the Facilitating Environment* (Madison, CT: International Universities Press, 1965).

[4]   J. Bowlby, *A Secure Base* (New York: Basic Books, 1988).

[5]   M. Meaney, "Nature, Nurture, and the Disunity of Knowledge," in B. McEwen, *The Unity of Knowledge* (New York: Annals of the New York Academy of Sciences, 2001).

and adult mutually regulate and scaffold their engagement with each other and the world by communicating their intentions and responding to them."[6] The development of meaning-making, the experience of biopsychological states of consciousness, and the engagement with others and the world are rooted in this first developmental attachment. The experience of synchronized connections in the mother-infant relationship lays a foundation for growth, exuberance, and the curiosity to explore and experiment with the world. Art-making grows out of these experiences and nonconsciously influences adult artist's creation of art and the viewer's social-emotional encounter with art.

CHILDREN'S ART: REGULATING SEPARATION ANXIETY

When the infant-mother relationship is disconnected, stress triggers flood the brain and body with emergency messages: danger! Separation anxiety in infants can show up as anger, irritability, sadness, and withdrawal. This fragile period of bonding is vulnerable to stress from many levels: within the child's health or temperament, the mother's mood state, stress on the infant-mother dyad from the environment, etc.

One of the ways mothers facilitate a secure bond is in play with their babies. As the mother provides opportunities for infant artistic expression in drawing, as she validates the baby's production, they cocreate shared images that secure their attachment.[7] Practicing visual motor skills, attentional focus, and emotional expression to create visual images in a shared, positive emotional experience, integrates all aspects of biological and social development. Implicit memories of these times making art with mother are nonconsciously stored in the baby's brain, laying a foundation for positive emotional value with art as an adult. Encounters with viewing art can trigger these positive emotions later in life, even though they will not be retrievable as memories.

As infants gain mobility and move into the toddler years, they gain enough autonomy to spend time alone. Separation anxiety triggered by distance from mother can be self-soothed with transitional objects and

---

6   E. Tronick, *The Neurobehavioral and Social-Emotional Development of Infants and Children* (New York: Norton, 2007).

7   L. Proulx, *Strengthening Emotional Ties through Parent-Child-Dyad Art Therapy* (Jessica Kingsley Publishers, 2002).

phenomena like teddy bears and blankets, and especially in creating art.[8] The creation of art provides practice for the expression and management of negative emotions of fear, anger, and confusion that accompany separations. Art-making and meaning-making intertwine for lifelong potential to self-regulate and overcome disconnections and adversity.

2.3 SOCIAL ANXIETY AND SHAME

Shortly after learning to babble *Mama* and *Dada*, children utter their first linguistic identity declaration, "No!" This monosyllabic exclamation begins the "terrible twos," and the beginning of learning shame, the core emotion of social anxiety. Mothers in all cultures struggle with securing their bond with their toddlers while establishing the hierarchy of social authority. "Time out" and "go to your room" sting in the moment of the reprimand, and ache in the forced isolation that disconnects the toddler from mother and family.

As budding storytellers in kindergarten, *social anxiety* predominates: "Did I say the wrong thing? Did I do something to upset someone? Did I tell a lie? Will I be punished?" The excruciating process of learning shame molds the neural networks for social anxiety. The chalk line between right and wrong, validation and shunning, is etched in the memories of the brain by the pain of shame if one is caught on the wrong side. "Am I accepted by others?" implicitly shapes the emotional experience of social encounters with parents, peers, and others. The foundation of a secure sense of oneself teeters on the internalized emotions and images that derive from social experience.

EVERY PICTURE TELLS A STORY: REGULATING SOCIAL ANXIETY

Children learn that "every picture tells a story" as they begin to scribble line drawings of people, places, and things. With their imagination, children discover the difference between "things in the world" and "things in the imagination." Art-making empowers the child's imagination to create a thing in the real world that represents the imaginary control over an important event, object, or situation. Rendering negative emotions and experiences in fixed, controlled images provides

---

[8]   D. W. Winnicott, *The Maturational Processes and the Facilitating Environment* (Madison, CT: International Universities Press, 1965).

the child with a sense of safety and security. When asked to describe in words what has been drawn or painted, children narrate their images in stories of personal meaning.

With imagination, children utilize art to distance themselves from the reality of external things.[9] In figure 5 a child draws a person leaping from the World Trade Center. (fig 5) Children can stay connected with positive emotions and experiences; they can flee from frightening or sad situations; they can gain control over terrifying experiences. The power of creative imagination for resilience to emotional and physical adversity begins in childhood.

We initiate children into our culture of images and words with pictures and stories. As children gaze at pictures of "The Three Little Pigs" or the "Grasshopper and the Ants" and listen to the allegorical tales about impulsivity, risk, self-centeredness, adversity, and planning for the future, the social lesson of right and wrong is coupled with the emotions of pride and shame. Allegorical pictures reflect the social lesson of survival through cooperation and connection with others.

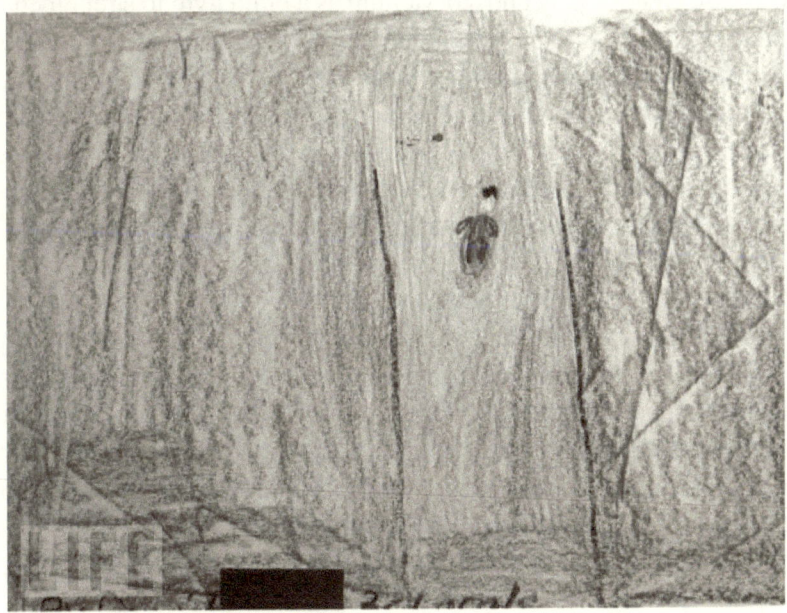

Fig 5 Child's art after 9/11/2001

_____

[9]   H. Krieger, *Children's Art*, ed. A. Tesche-Mentzen and H. Koelbl (Munich: Frederking and Thaler, 2003).

The developing brain fuses and records words and images. In parental storytelling gestures and mime, in bedtime reading with picture books, and in family watching of TV and movies, words and images are seamlessly sewn into the neural fabric of emotional memory. Fear and anger, sadness and joy, surprise and disgust are safely felt with pictures and stories, fairy tales and fables. As our brain's appetite for stories and images develops greater cognitive and emotional complexity, the developing brain wants to taste everything from Shakespeare and Rembrandt to B movies and sexy computer games. Each encounter with pictures that tell stories gives our brains practice mastering our negative emotions and attaining our positive emotions. Beginning in early childhood and arcing across the entire life span, art provides the brain with wordless stories of all kinds that facilitate social learning and empower regulation of social anxiety.

Children learn social intelligence[10] by practicing empathy, and the interactive skills of sharing, turn-taking, negotiating, and compromising. In ambiguous social situations, the social brain gets creative with half-truths, minimizing emotion, bending and exaggerating events to suit the situation. Children use creative play and storytelling as a brain gym to strengthen their imagination, to self-soothe when they are alone, to engage an audience, and to practice sharing and taking turns with peers. The brain's power for creating narratives in play, music, and language provides the core for adaptive resilience to adversity and the first set of skills for making art, music, dance, and drama. Art links children with positive emotional ties in families, to enhance eye-hand-body motor coordination, and to transmit culture.

2.4 IDENTITY AND GENDER ANXIETY

The recent cultural acknowledgement of transgender persons demonstrates a willingness to accept nature's diversity and demystifies the myth of the gender binary found in most religious dogma. Identity and gender identity are complex processes rooted in biological and social development. The cultural pressure to conform to "should-be-this—way-or-that-way" stereotypes affects all children and their families as they struggle to conform to social expectations. Emerging in the first two years of life, the gender identity of a toddler is one of the first meeting grounds of biology and culture. The limbic brain's capacity for creating retrievable memories (explicit memory) begins at about two years old

---

[10]   D. Goleman, *Social Intelligence* (New York: Bantam, 2006).

when the hippocampus is mature enough to perform this function. It is at this time that the story of "I am a boy or girl" begins, and the emotional struggle to "fit in" ensues—identity and gender identity anxiety begin here, simultaneously intertwined with social anxiety and the capacity to feel shame. A sense of "self" takes shape during childhood, largely defined by parents and culture. It must deal with everyday adversities to gain a sense of permanence—"I can catch a ball well; I feel pride and the envy of others so I can be a baseball player"—the self builds its own history through actions and relationships.

IDENTITY AMBIGUITY IN ART

"When in Rome, do as the Romans." From the brain's point of view, identity is a series of computations that coordinate the behavior and attitude of the self for optimal adaptation that provides a sense of safety and security to the context it's in. Identity is by nature a process of adaptation more than an unchangeable fixed state. Social conformity makes demands on both behavior and attitude while it provides an identity with the safety of belonging to a group—distinguishing oneself from others in the group is a delicate balance of individuation and conformity.

Artists have long had the reputation of being "outsiders" of mainstream society because their behavior and talent, making art, has often been seen as special or eccentric. To make authentic art requires a connection with the True self, the nonconformist self. The False self is the conformist self that is poorly equipped for art-making, but can make non-original designs that are pleasing to others.

Identity and gender identity have been explicit and implicit subjects in art. The theory that Leonardo's famous portrait, *Mona Lisa*, (fig. 6) is actually a self-portrait may have inspired Marcel Duchamp's self-portrait as Mona Lisa with a mustache painted on it to reveal the charade. Duchamp, who had already created art with self-portraits as Rrose Sélavy, a woman, implicitly brings to the viewer the connection of Leonardo's identity as a possible transvestite or homosexual. Throughout history, people who do not conform to the gender binary have been discriminated against, requiring them to "hide in the closet" and assume a False self to meet the public's expectations. The fact that society values the function of artists and tolerates their nonconformity makes the arts a natural refuge for anyone feeling that their True self is not accepted by mainstream society.

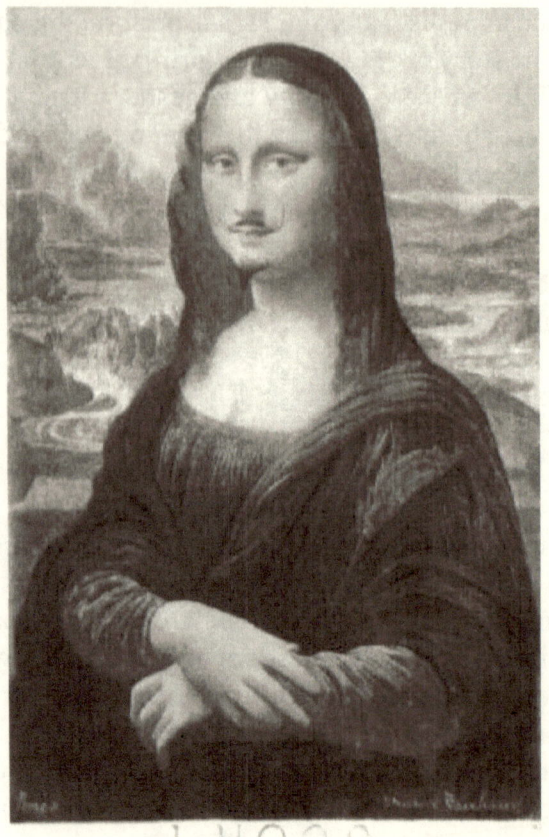

Fig 6 Marcel Duchamp *Mona Lisa*

2.5 EXISTENTIAL ANXIETY: FRANKENTEEN OR HUMAN BEING?

Existential anxiety emerges as our bodies make the stressful transformations of puberty and the trillions of brain connections begin to perform the cognitive acrobatics of abstract thinking and reflective consciousness. Pubertal teens often feel like ugly ducklings or "Frankenteens." They can create their own personal hell with existential questions: "Who am I? Where do I come from? What am I doing here? What does it all mean? Does anyone really know or care who I am?"

During these years, teens can stumble into a pit of existential anxiety where they terrify themselves with their imagination, taking their budding identities hostage in a "no exit" world of self-doubt and social insecurity. Their brain's naive computations often conclude that people, places, and things are either an idealized "all-good" or devalued "all-bad."

Separation and social anxieties are amplified and compounded by the brain's ability to scare itself with imagination. Identity insecurity is triggered by physical appearance, social acceptance with peers, gender and sex, intelligence, and physical prowess. The teen brain can be quickly swept into a conundrum of conscious and nonconscious emotions that drive impulsive and risky behaviors. Adolescents test physical, emotional, and social limits to discover their identity boundaries. Powerful sexual and aggressive urges present the developing brain with enormous social-emotional conflict. The mastery of these feelings and urges is critically dependent on the social context and positive emotional connectedness with family and society. The teen brain's new cognitive powers can embark on philosophical discourses, create complex meaning with art and poetry, explore higher mathematics, and enter political maneuvers with peers and authorities.

TEEN BRAIN GROWTH AND SYNAPTIC PRUNING

Beginning with puberty, the brain's cortex, gray matter, undergoes an explosive increase in the number of neurons and their synapses (the chemical-electrical connections between neurons that conduct the information that produce neural network activity and consciousness). During the midteen years, these excessive synaptic connections are "pruned" by usage. If a neuron's synapses are not engaged in activity, it will die. "Use it or lose it." Synaptic pruning is directly related to neural activity from experience. The teen's life experience literally molds the neural networks that shape the person he/she will become.

The plasticity exhibited by the brain during pubertal development makes individuals more vulnerable to perturbations and stress. Stress-induced neuronal remodeling during this profound period of plasticity may affect the teenager's stress responsiveness, emotional behavior, and cognitive behavior at the crossroads of dysfunction and creative resilience.[11] Teenagers are much more likely to develop addictions to drugs and alcohol than young adults due to their stage of brain maturation.[12]

Despite the possible vulnerabilities of the pubertal brain to stress, adolescence may provide opportunities to alleviate adverse effects of

[11]   B. McEwen and R. Romeo, "Stress and the Adolescent Brain," in *Resilience in Children*, ed. B. McEwen et al. (New York: Annals of the New York Academy of Science, 2006).

[12]   N. Volkow, "Drug Addiction: A Brain Developmental Disorder," in *Resilience in Children*, ed. B. McEwen et al. (New York: Annals of the New York Academy of Science, 2006).

stress experienced earlier in development. As synaptic pruning continues until the early twenties, the brain can launch its exploration and discovery of personal meaning and kindle the creation of an authentic identity. Art provides the maturing social brain with experience that empowers the skills to modulate stress and secure the reciprocal connections between individuals and culture.

ART, REFLECTIVE ATTENTION, AND ABSTRACT THINKING: REGULATING EXISTENTIAL ANXIETY

Adolescents develop the adaptive tools for abstract thinking and reflective attention to explore conflicts and make decisions about possible futures, what has been called 'formal operations'[13]. Using these tools of imagination for future planning empowers teens to delay gratification, regulate their emotions, connect socially, and pursue their dreams to create personal meaning. Recent neuroscience demonstrates that repeated activation of the brain's attention networks increases their efficiency. Neuroimaging studies have also proved that the following specialized neural networks underlie various aspects of attention. [14] (Fig 7)

- the alerting network, which enables the brain to achieve and maintain an alert state
- the orienting network, which keeps the brain attuned to external events in the environment
- the executive attention network, which helps us control our emotions and chose among conflicting thoughts in order to focus on goals over long periods of time

[13]   Piaget, J. *The Psychology of the Child*. New York: Basic Books, 1969.

[14]   Posner, M. *How Art Training Improves Attention and Cognition in Cerebrum*. New York: The Dana Foundation, 2010

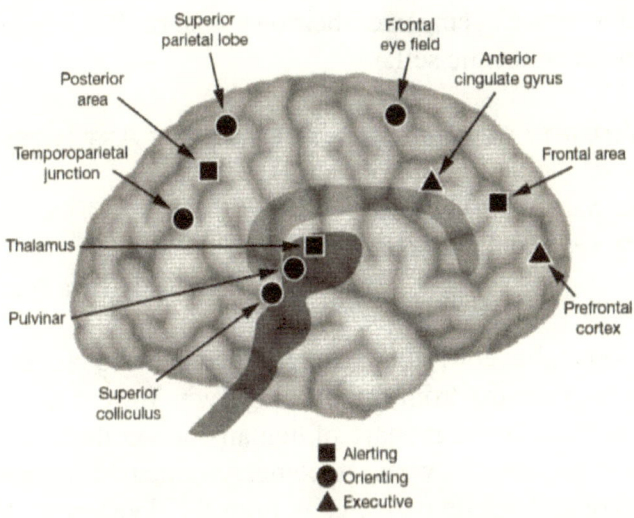

Fig. 7. Brain networks that underlie different aspects of attention include the alerting network, the orienting network and the executive attention network. Arts learning may contribute to improved cognition by improving the efficiency of the executive attention network. (Courtesy of M. Posner.)

Practice using one's imagination integrates emotion, cognition, and motivation to critically support the development of a resilient, authentic sense of self. Drawing, painting, writing, acting, dancing, musical performance enhance the integration of body, mind, and identity.

The adaptive value of finding personal meaning through expressive arts cannot be overestimated. Through the process of making art and by observing art, the brain practices becoming a conscious, authentic self. The mental and emotional experimentation involved in the creation of art gives the "synaptic self"[15] a workout. The adaptive regulation of existential questions "who? what?, where?, why?, how?' are empowered with art. With increased power of cognitive and motor control, teenagers can embark on the practice of expressing complex emotions with art and reflecting on personally meaningful interpretations of art. An authentic sense of self emerges with successful use of reflective attention and abstract problem solving. To regulate emotions and impulsivity, teens need many trials and failures and retrials of pursuing wishes and overcoming adversity physically, emotionally, cognitively, and socially.

---

[15]   J. LeDoux, *Synaptic Self* (New York: Viking, 2002).

By utilizing the expressive powers in art and reflecting on the art of others, adolescents can empower their own goal of developing personal meaning and an authentic self.

2.6 The Anxiety of Impermanence and the Art Imperative

What thou lov'st well shall not be reft from thee; What thou lov'st well is thy true heritage.
—*Pisan Cantos* (LXXXI), Ezra Pound

On 11 September 2001, the world watched in horror as people chose to jump to their deaths from the burning Twin Towers rather than be consumed by fire. The nightmare of humankind's collective terror was mirrored over and over for weeks to come. And then the stories emerged from the recorded mobile phone calls from the doomed. "I love you," was the final message heard again and again.

When we face our doom, the meaning of our final gestures and words seems to reach across the inevitability of impermanence to the living, as if to transmit a timeless message of love. "I may be gone, but my love for you and for life lives on in the memories of all that knew me." This powerful legacy challenges the fears and sadness that so spontaneously arise at the time of death.

I use the term, *anxiety of impermanence*, to indicate the perpetual invitation to human consciousness to ask, "What if . . . ?" Separation, social, and existential anxiety commingle to become the anxiety of impermanence, a pervasive malady of humankind. In the Four Noble Truths, Buddha observed that

1. Life is suffering.
2. The cause for suffering arises from our clinging attachment to things and ignorance of the impermanent nature of reality.
3. The cessation of suffering comes with understanding that the essential nature of things, feelings, and thoughts is emptiness, devoid of permanent meaning. Acceptance of the truth of the impermanent and relative nature of our perceptions opens the way toward a deep understanding of emptiness.

4. The liberation from our suffering state of consciousness arises with the direct, intuitive experience of emptiness through mental disciplines of attentional regulation.[16]

We are wired to survive today and to learn how to worry about tomorrow. "What if Mommy abandons me? What if I do something to embarrass myself? What if I fail to figure out what to do with my life? What if I don't get into the right college or get the good job? What if I don't meet the right mate? What happens to me when I die?"

We are also wired to connect with each other. Our connections within our selves, between each, and with meaningful cultural values are secured with our creative potential. Internal links that regulate mind and body and social-emotional connectedness form a powerful antidote to the three developmental anxieties that leave us fretting in a quandary of doubt. Evolution has empowered our brain with the capacity to develop creative problem-solving skills at each level of developmental stress. The primary goal in creative problem solving is to form secure connections with our body and identity, with others, and with the world.

CONTEXT AND CREATIVITY: RATS, CATS, AND SHARKS/ SAFE OR DANGEROUS?

The Physical Impossibility of Death in the Mind of Someone Living by Damien Hirst (1991) (fig. 8) presents a gigantic shark suspended in a glass vitrine filled with formaldehyde as "art."

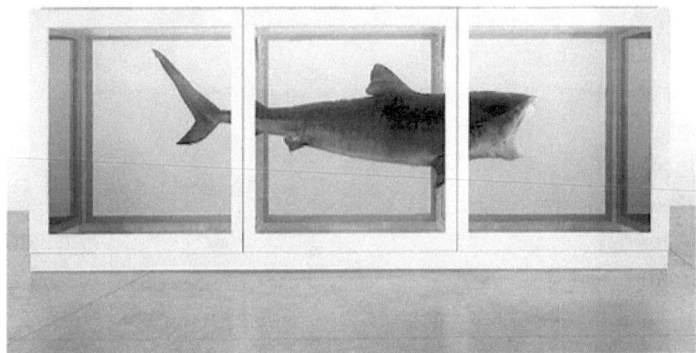

Fig 8 Damien Hirst The *Physical Impossibility Of Death in the Mind of Someone Living*

---

[16]   Dalai Lama, *The Four Noble Truths* (London: Thorsons, 1997).

My first reaction on seeing it at the Venice Biennale was "I'm in the wrong museum." In the context of a Natural History Museum, this object would be "just another dead animal" labeled under its name in Latin for genus and species. But Hirst creates a challenging visual metaphor that confronts the viewer with an opportunity to reflect on their own "anxiety of impermanence" by placing the shark in the context of an art museum. Hirst triggers the allostatic system with the unexpected shark and engages the viewers' minds and emotions as much as their eyes. At many levels of meaning, the brain must resolve this emotional encounter with art.

The second association I had was triggered by a startle response when I caught a glimpse of my reflection as I looked into shark's mouth through the blue tinted vitrine. "I'm glad it's dead and I'm not in the water looking at it!" I thought, reflecting on the safety provided the context of art.

Like Andy Warhol's Campbell's Soup paintings, or Marcel Duchamp's infamous Urinal, Hirst plays with known objects by placing them in an art context to trigger complex associations and emotional responses in the viewer. Picasso juxtaposed a bicycle seat and handle bar to create a bull's head (fig. 9) utilizing ordinary "found objects" and transforming them into art objects. Artists juggle objects, confound context, and manipulate meaning to startle our expectations, challenge our preconceptions, and invite reflection. René Magritte painted a tobacco pipe and named it "This is not a pipe"! The implicit message of Consilient Art is, "this is not what it looks like; look closer; reflect; there's more here than meets the eye."

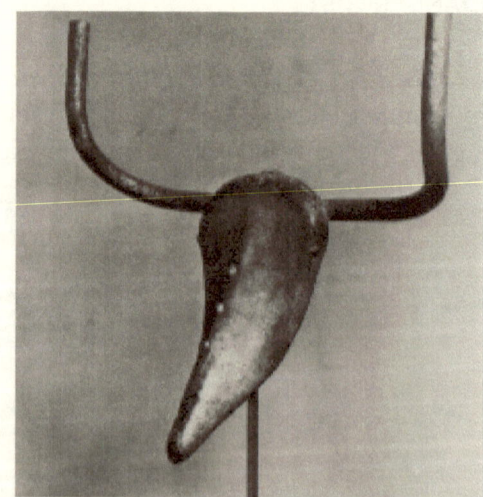

Fig 9 Picasso *Bull*

When laboratory rats are placed in a strange but safe context with other rats that are at home with toys, food, and shelter, they will explore, socialize, and begin to play. Placed in the same context that has had a few drops of cat urine placed in a corner will cause the rat to freeze, seek shelter, isolate from others, and fail to explore or play. So powerful is the olfactory cue (the scent of cat urine) that it totally overrides the enriched environment, shutting down the rat's natural urges to adapt to the context.

Art confronts us with contextual challenges that trigger our stress circuits in the safe context of an art gallery or museum or garden. From El Greco's elongated religious figures to Picasso's cubist portraits, artists push perception, challenged expectation, and invite reflection. Art engages heightened attentional focus emotionally, cognitively, and motivationally.

ART, AESTHETICS, AND THE IRRESISTIBLE INVITATION TO REFLECTIVE ATTENTION

Art exists somewhere between the internal, private world of emotion, thought, and memory and the external, public world of people, places, and things. Art bridges internal and external realities, and links the artist and the viewer consciously and nonconsciously. Art fixes a marker that transcends time by mirroring our primordial past, our present moment, and in kindling contemplation about our possible futures. Art triggers an automatic urge to explore the art object, providing an irresistible invitation to reflective attention.

Aesthetics emerge from a matrix of connections between brain, body, identity, relationships, culture, and the world. The philosopher Alexander Gottlieb Baumgarten coined the term *aesthetics* in 1735 to mean "the science of how things are known via the senses."[17] Art offers both artist and public a reflective experience on their external, public realities and their internal, private realities. Art is explicitly exhibitionistic and implicitly encrypted. The experience of art and the attention to aesthetics encompass at least three perceptual and attentional domains: the material present, the private ephemeral emotions of artist and audience, and the transient meanings of art that emerge in the viewer's experience. Art presents an irresistible invitation to the viewer: "Take a look, check it out, explore and reflect!"

Blockbuster shows of Tutankhamen (fig. 10) that tour the world like rock stars or the sensational exhibition of Damien Hirst's $100,000,000 diamond skull (fig. 11) are designed for shock and awe in the material art business of the present—they also implicitly remind us of our impermanent nature and the beauty we create while being alive. The

---

[17]   Peter Kivy, ed., *The Blackwell Guide to Aesthetics* (2004).

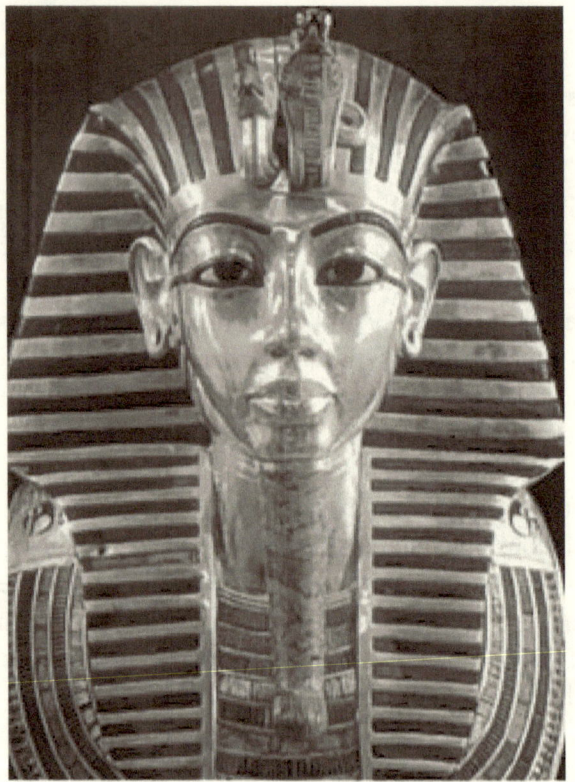

Fig 10 Tutankamen Death Mask

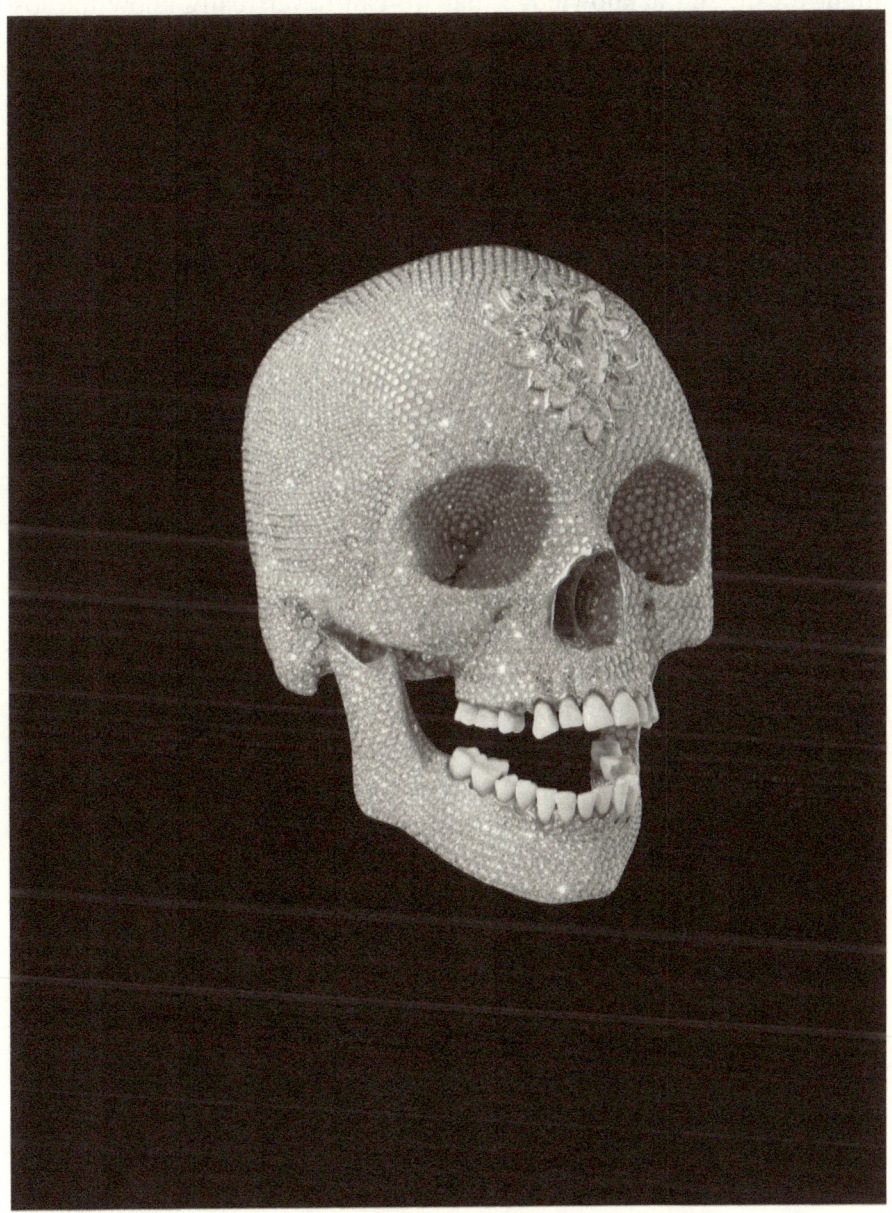

Fig 11 Damien Hirst *For the Love of God*

ephemeral emotions of the artist and the viewer are first and foremost private experiences. The timeless flow of transient meaningfulness of art can be glimpsed when we think of what the Paleolithic cave paintings might mean, and acknowledge that we can never really know the lost personal meaning of the cave painters and their tribe. Art connects us to ourselves, to others, to the world, and to our primordial roots.

The brain's capacity for paying attention requires effort to focus consciousness on a thing, a thought, a feeling, a sensation, or an intention. Different from focused attention, perception can be conscious and nonconscious. Perception of external reality becomes conscious through the brain's reflective mechanisms for effortful attention. Each night, when external reality disappears, the brain creates an internal holographic world during dream sleep that becomes our new reality. Emotionally intense dreams trigger our allostatic system to up-regulate our heart rate and breathing. Dreams plunge us into a full-spectrum juxtaposition of sensory experience. Our brain's dream apparatus can transport us in time to the most remote reaches of our memory or imprison us in the drudgery of our day-to-day experiences. Dreams can subjugate us to our darkest fears or free us into the ecstasy of our wildest fantasies.

The perception and meaning of both our internal world and the external reality depend on our attention. When an emotionally charged dream sticks in our waking consciousness, we would do well to pay attention to it. By reflecting on the dream, we can interpret the message that we have just received from ourselves.

If we are driving and don't pay attention to the stop sign, we run the risk of an accident. When a here-and-now event startles us into special awareness, we would do well to pay attention to it. By reflecting on the event, we can interpret the experience that we have just had from the "thrownness"[18] of everyday life.

We can be triggered to fight or flee, feed or fornicate by external or internal perceptions. When we cultivate our attention with effortful practice, we open up new ways to deal with instinctual urges. When emotions without reflection drive action, we tend towards self-serving behaviors that are often self-sabotaging or destructive. When we chose to refocus our attention toward our wishes, reflection kindles creative problem solving. Resilience and adaptation are direct results of effortful reflective attention.

---

[18]   Martin Heidegger, *Being and Time*, trans. John Macquarrie and Edward Robinson (New York: Harper & Row, 1962).

Infants begin practicing attentional focus with all of their sensory experiences: sight, sound, taste, touch, and smell. The infant-mother relationship is the first "attentional training studio" for the developing brain. The mother's attunement to her baby facilitates joint attention, the foundation of social intelligence and a key to optimal brain development. The maturational processes and neuroplasticity of the brain (the brain's ability to rewire itself from experiential learning) allow for an ever-increasing ability to make critical attentional distinctions throughout life. Our taste for new foods or more-refined flavors can be cultivated with explorations in cuisine. Our appreciation and emotional openness to new music and foreign languages emerge with listening. Looking at art, architecture, fashion, design, from ancient times, foreign cultures, and contemporary artists challenges our fixed sense of these domains.

The scaffolding of exploration, experimentation, and creative expression from infancy to toddler to teen prepares us for the ultimate test of adaptive resilience and the mastery of the anxiety of impermanence.

WIRED FOR ART—ATTUNEMENT, MIRROR NEURONS, AND MINDFULNESS

Art, language, social-emotional connectedness, and culture have always been separate domains of knowledge until very recently. Giacomo Rizzolatti's recent discovery of mirror neurons in monkeys is forever changing the way we think about all of these academic disciplines. When Rizzolatti was studying single brain cells in the motor cortex that become activated with movement—picking up a peanut—he discovered that certain cells also became activated when the monkey saw a man pick up a peanut. The cell "mirrored" the action it saw—he called them "mirror neurons." Monkey see, monkey do! Our brain has specially designed cells that learn by imitation of intention.

The implications of this discovery offer new explanatory power for many of the mysterious attributes that we have evolved to become humankind. Mirror properties have been discovered in the human brain in what has been called the Mirror Neuron System. This system creates representations of other's actions, intentions, and emotional states. Through mirroring, we learn language, gestures, and meaning.[19]

---

[19]   M. Stamenov and V. Gallese, *Mirror Neurons and the Evolution of Brain and Language* (Amsterdam: John Benjamin's Publishing Company, 2002).

That the mirror neuron system allows for attunement to the intentional state of others demonstrates the profoundly social nature of our brains.[20] With attunement, we can develop emotional resonance, empathy, and coordinate joint attention with others for creative actions. When this attunement to others is reflected inward, we can develop a deep sense of attunement to our own mental and emotional process. With this reflective observation of our own mind, we can experience our essential nature.

Cultures throughout history have devoted special attention to mental reflection. Mind training helps overcome the universal mental stress of humankind, what I call the anxiety of impermanence. We have created meditation, prayer, yoga, tai chi, and many other attentional disciplines to cultivate mindfulness. Mindfulness involves focusing our attention on the here and now, our inner mental-emotional process, and attunement to our intentions. The many benefits of mindfulness training include reduction in stress, immune support, heightened awareness of the moment, and a deep sense of peaceful acceptance of our impermanence. Spontaneous compassion for oneself and others springs from this knowledge of our essential nature.

Thousands of years before we used our reflective attention to explore consciousness and create attentional training methods, our brain's mirror neuron system evolved the skill to create images and words that connect us with others. These social-emotional connections calm us and teach us the ways of our elders. With abstract images that represent these connections and words, we can communicate our intentions to others and coordinate our adaptive efforts with groups across time, long after we are gone. Each generation can leave a pictorial trace of their experience to connect with future generations. The cave paintings of Lascaux and Chauvet command our attention for complex reasons, among the obvious being their timeless beauty and their age (15-30,000 years old). These images mirror the lives of those gone before us by condensing intention, meaning, and charged emotions for our reflection. By mirroring the brevity of our lives and reflecting the timeless connection to our ancestors and future generations, art has helped us master the anxiety of impermanence for thousands of millennia.

---

[20]   D. Siegel, *The Mindful Brain* (New York: Norton, 2007).

REFLECTIONS ON THE ART IMPERATIVE

A consilient approach to art reveals humankind to be endowed with an automatic "art imperative" that scaffolds language, social-emotional connectedness, and authentic identity development. Art plays a key role for the integration of these attributes for individual and cultural resilience to adversity. As the structure of societies, governments, and religions emerge and disappear over the millennia, art endures, transforming itself, reinventing itself, but never ending. We make pictures, objects, music, and dance that tell stories about our lives in the past, ourselves today, and the future of our species. Art functions as a never-ending story that provides meaningful sanctuary across the boundaries of culture, time, and space for our True selves and for others.

PART II

ART AND RESILIENCE

It is not the strongest of the species that survives, nor the most intelligent that survives. It is the one that is the most adaptable to change.

—**Charles Darwin**

Chapter 3.0

THE LEGACY OF LASCAUX: THE SCAFFOLD OF ART AND CULTURE

In past times when one lived in contact with nature,
abstraction was easy; it was done unconsciously. Now in
our denaturalized age abstraction becomes an effort.
                                        —Piet Mondrian

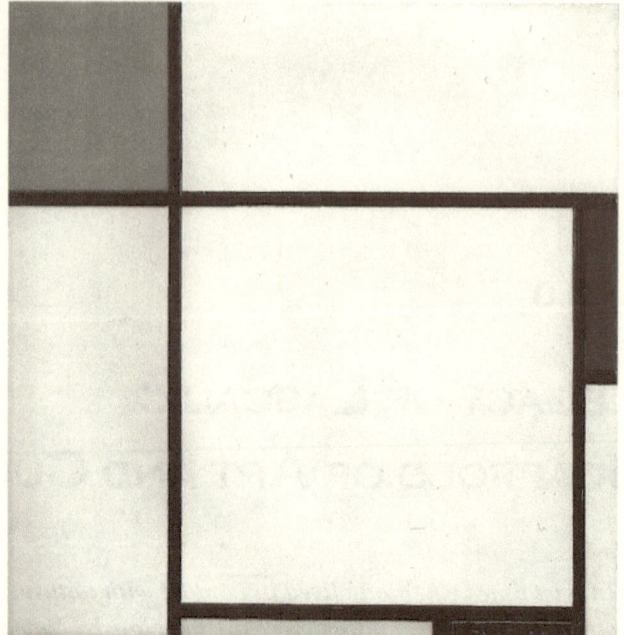

Fig 12 Piet Mondrian

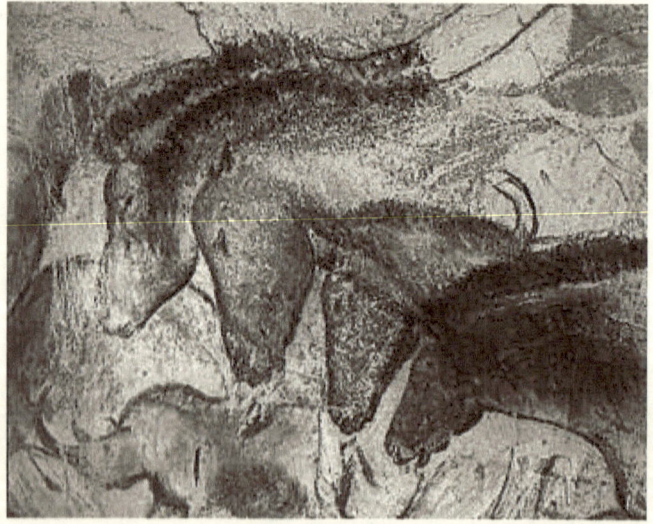

Fig 13 Chauvet

3.1 ART WINS: NEANDERTHAL VS. CRO-MAGNON AND THE GREAT LEAP FORWARD

The cave paintings of Cro-Magnon (European early modern humans) at Lascaux, Altamira, Chauvet (fig. 13), and others date back some 30,000-40,000 years. Art-making is evidence of a "great leap forward"[21] in the evolution of complex cognitive-social functioning—that is human culture. Decorative objects from shells and beads to tools were created by human beings for thousands of years before the emergence of the paintings—nothing like these paintings ever existed before and the awe they inspire goes far beyond the pleasures taken with decorated objects or painted clothing.

In roughly the same period of time that these paintings emerged, Neanderthal, Cro-Magnon's rival for food, became extinct—the Neanderthal extinction hypothesis cites many factors that contributed to their demise, especially the competition with Cro-Magnon. Several theories cite the competitive edge that Cro-Magnon's cognitive-social superiority may have empowered their capacity to exterminate Neanderthal over time. The focus of this chapter is the imperative role that art-making plays in the evolution of cognitive-social complexity and culture.

3.2 ART, LANGUAGE, AND THE BRAIN

Darwin said,

> In the long history of humankind (and animal kind, too) those who learned to collaborate and improvise most effectively have prevailed.

The emergence of language and art-making mark the two most distinguishing features of modern human culture that have been critical to its prevailing over all other species. Language facilitates collaboration—the coordination of coordinated actions,[22] and art provides a virtual world for abstract improvisation—both are grounded in symbolic and abstract cognition that emerge from the evolved brain's neural architecture for

---

[21]   J. Diamond, *The Third Chimpanzee* (New York: Harper Perennial. 1992).

[22]   H. Maturana and F. Varela, *The Tree of Knowledge* (Boston: Shambala. 1987).

organizing and communicating complex information. The human brain processes information and experience from life and consolidates it into abstract symbols charged with many different levels of meaning that offer myriad interpretations, associations, innovations, and creativity—the brain can observe a simple stone and, using the imagination, construct a tool, a weapon, or a sculpture from it—this capacity for visualizing many possible uses for the stone is one of the hallmarks of how the human mind can alter the environment to create culture.

Language is constructed with words, meaningful units that have a meaning all their own even before they become part of a sentence—naming people, places, things, actions, feelings, etc. The vocabulary of language consists of words with more or less fixed meaning that together are combined through syntax to form additional meanings. Language used for daily communication is largely an explicit form of communication—poetry being a form of language usage for art that uses ambiguity to expand possibilities of meaning.

Language is largely a functional tool for the coordination of social activities or reflection on shared history or emotional experience, but not a tool for contemplation—talking to oneself out loud may be idiosyncratic but not socially functional. With the emergence of reading and writing of language came the possibility for private learning and contemplation of words and stories—closer to the "viewing-contemplation-interpretation" that art can inspire. The evolution of writing systems was preceded by picture and symbol making. Both language and art inspire the production of the other—Keats's *Ode to a Grecian Urn,* the world of art criticism, and the entirety of religious art inspired by the Bible are examples.

The units of art are neither fully defined, nor do they have a fixed meaning. A brushstroke or daub of orange paint, when seen in isolation, outside of the art, has no special meaning, unlike words. The "Combine Paintings" of Robert Rauschenberg are conglomerations of found objects ranging from lightbulbs to newspaper to cardboard boxes to taxidermy animals, etc. used as compositional elements for their color, texture, and even their functional meaning outside of art as a symbol or metaphor—Rauschenberg makes it clear that anything can be used in the creation of art (fig. 14).

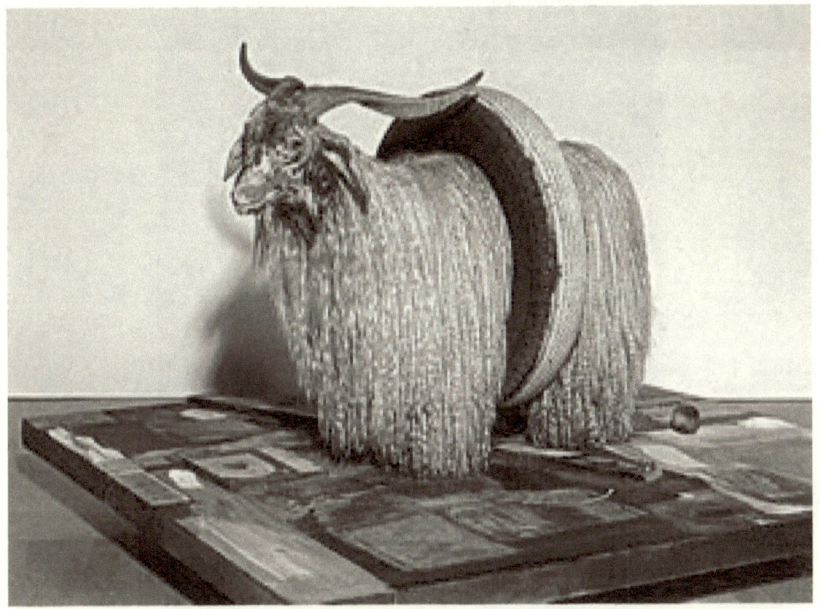

Fig 14 Rauschenburg Monogram

Artworks contain more ambiguities than language. Although the "picture" element of art can be explicit, the unique stylistic presentation or composition creates an expanse of possible meanings open for interpretation and discussion—a portrait by Ingres inspires very different cognitive and emotional responses than a portrait by Van Gogh or Picasso.

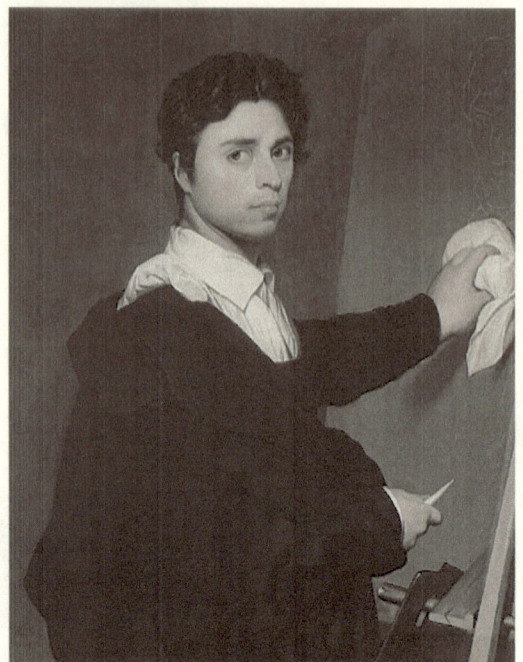

Fig 15 Ingres Self-Portrait

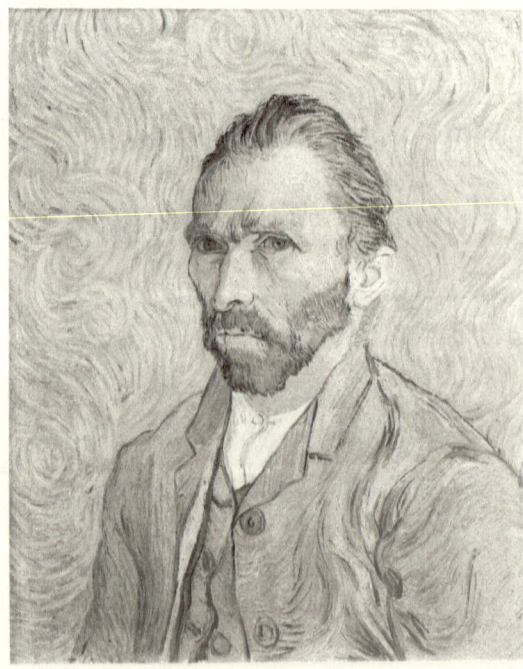

Fig 16 Van Gogh Self-Portrait

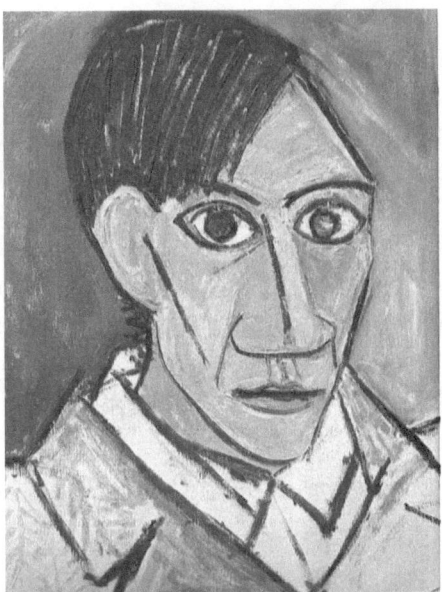

Fig 17 Picasso Self-Portrait

Art has no apparent functional use (in East Asia, walls and doors have often been employed as a ground for painting), but it is intended for both private contemplation and mass viewing. Art promotes intellectual discourse, invites multiple interpretations, and triggers varied emotional responses. The relative lack of precision in the symbolic nature of art renders it a powerful communicative system.

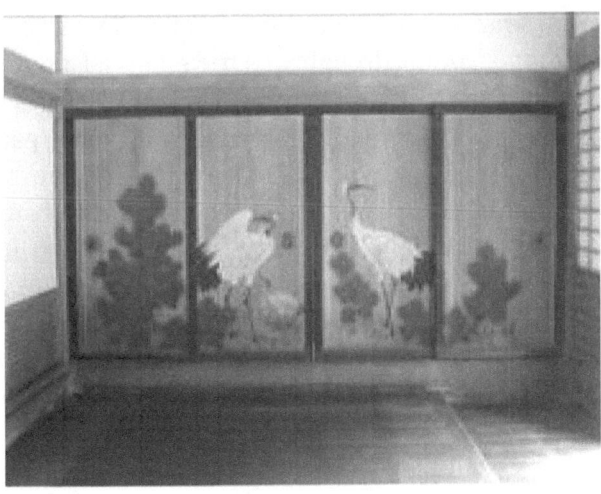

Fig 18 Kongoblji Shrine, Japan

Brain mapping demonstrates the location of language production and comprehension to be dominantly in the left hemisphere of the brain.[23] Similar attempts to map the centers for art production reveal that many areas of the brain contribute to art-making and art comprehension—art-making and aesthetic appreciation of art are nonlocalized in brain mapping with specialized areas, like the visual cortex, contributing to the visual creation and interpretational aspects of art. Art-making capacity in artists survives most types of localized brain damage, and even in degenerative dementias, art-making survives for long periods of time. There are even dementia cases where art-making emerged in nonartists after the dementia has progressed—speculation that the degenerative frontal lobes that suppressed artistic expression failed and the urge to make art was disinhibited!

3.3 MERLIN DONALD'S THEORY: THE EVOLUTION OF COGNITION AND CULTURE

Homo sapiens originated in Africa, based on both genetic and fossil evidence. They first migrated to Europe and Asia about 100,000 years ago, and a second migration occurred about 60,000 years ago, spreading throughout the world. Archaeological evidence reveals the use of red ochre pigments and shell ornaments with both the first and second migrations, implying symbolic cognition.

The theory is that the *Homo sapiens* who remained in Africa after the first migration underwent further social and crucial neurochemical and neuroanatomical brain changes that were adaptive as survival strategies in the face of a long-standing drought in Africa. This adaptive response of humans to the climatic changes contributed to technological innovations and improvements in tools as well as behavioral changes, particularly in relation to social grouping. The confluence of these factors could have reorganized cognition in the direction of symbolic and abstract thought that shaped the brain's capacity to produce art—the Art Imperative—that is a highly adaptive attribute facilitating human cultural resilience up to the present day.

Evolutionary psychologist Merlin Donald proposes a developmental scaffold to answer the question, "What transpired in the evolution of humans to make us who we are today?" His answer to this question

---

[23]   D. Zaidel, *Neuropsychology of Art* (Hove and New York: Psychology Press. 2005).

identifies a three-stage cascade of cognitive and cultural evolution: (1) Mimetic culture (+/ -2 million years ago) (2) Mythic culture (+/ -150,000 years ago) and (3) Theoretic culture (2,000 years ago). The emergence of Mimetic culture provides a foundation for the increasing complexity of Mythic culture, and likewise, the Mimetic-Mythic scaffolding is a springboard to the present-day Theoretic culture.

Mimetic culture is pre-verbal—it entails gesturing, dance, pantomime, visual analogy, and ritual—these elements involved action-metaphor, pre-symbolic and pre-abstract behaviors that allowed for tool-making, fire-tending, and the foundation of ritual behavior. Gestures as signs of communication occur in many animal species (fig. 19) and persist in rituals of Australian aborigine (fig. 20). This stage laid a

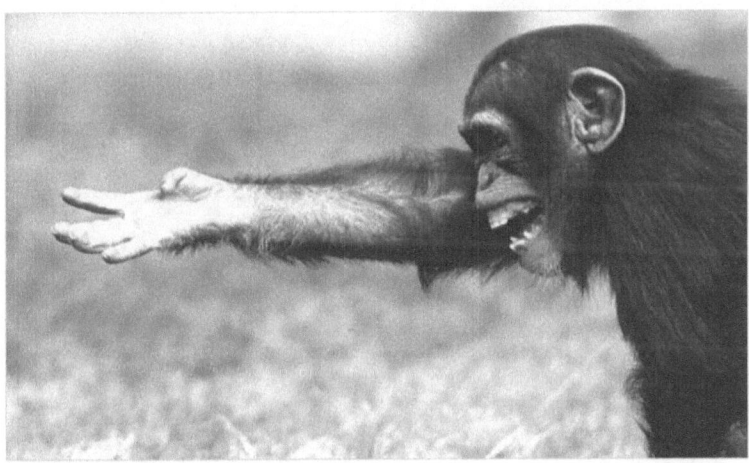

Fig 19 Chimp gesture

foundation for the emergence of spoken language—the encoding of action-metaphors into phonemic utterances. We see the remnants of mimetic culture in the gesturing of the pre-verbal toddler who points to the cookie jar and references his mouth when he is hungry. Sankai Juku, a contemporary Butoh Dance troupe from Japan, aims to connect performers and audience with the primordial Mimetic culture using body painting, mime, gesture, body language, costume, makeup, ritual, and otherworldly music (fig. 20)

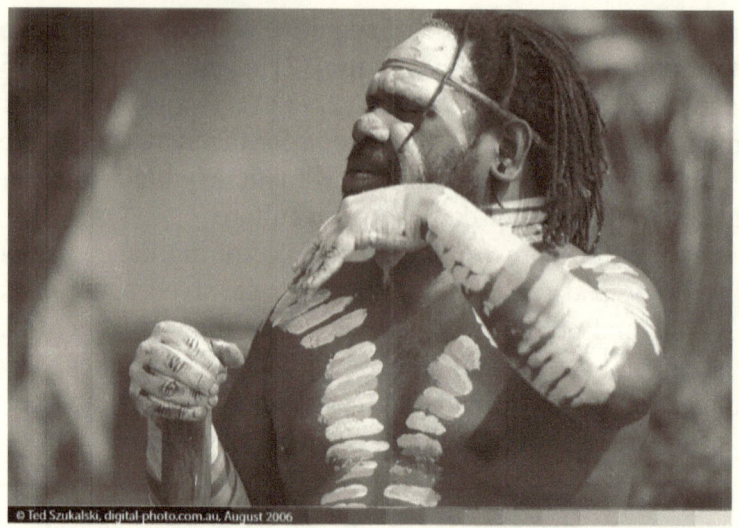

Fig 20 Aboriginal-Hands-Gesture

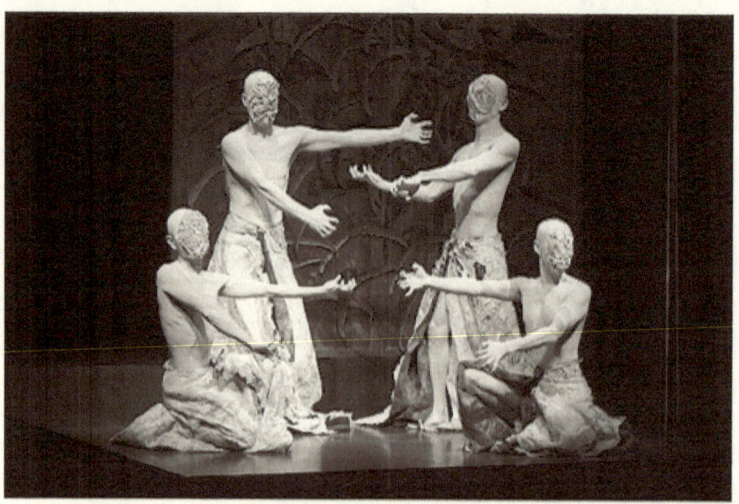

Fig 20 Sankai Juku

Mythic culture emerges with spoken language and its power to generate storytelling—the medium of storytelling allows for shared experience over time. Storytelling defies the limits of a single generation's lifetime by providing a story told and retold until it acquires the power of timelessness or myth. The myth contains archetypal stories of origin, meaning, and the future that become a potent remedy to the anxiety of impermanence so acutely felt by our prehistoric ancestors. They can also preserve notions of authority, gender, and morality. Mythic culture retains elements of Mimetic culture in rituals, costume, and gesture, which shape the art of the culture and persist in the practices of all religions today.

Theoretic culture is based on symbols, traceable in art-cultural history, and only recently emerged with the advent of sophisticated writing technologies and scientific instruments. The critical distinctions that identify Theoretic culture include bureaucracy, logic, and external memory devices such as writing, codices, mathematical notations, scientific instruments, books, records, and computers. The culture is organized by government, science, and technology that give rise to a diversity of art forms. Donald explains,

> In a global context, relative to the influence of Mimetic and Mythic domains, Theoretic culture is still a minority culture. However, it is disproportionately influential because of its place in the distributed cognitive systems that determine such things as our collective representations of the past and our tribal and class identities. Of necessity, even Theoretic institutions retain a Mimetic and Mythic element; human society cannot function without these more basic forms of representation, which carry out specific kinds of cognitive work.

Contemporary art includes expressive forms from Mimetic, Mythic, and Theoretic culture—art persists in every new generation just as language does. Art carries the legacy of past art forms that become integrated with present day issues. Donald declares that

> the new is always and inevitably scaffolded on the old, and as a result, art is ultimately a reflection of the deepest and most ancient form of human expression, mimesis.

3.4 FROM CHAOS TO CREATIVITY: THE PHOENIX PHENOMENON

The human brain is hardwired for survival—freeze, fight, flight—fear and anger being critical to the process. Social connectedness empowers human resilience to adversity, and when adversity and chaos overwhelm the brain and the freeze response becomes chronic, social disconnections can occur, and there is little hope for survival. Humans have survived because of their capacity for creative adaptation to the most harsh and extreme of adversities. In the face of chaos and confusion, the brain is somehow able to refocus, reflect, and reconnect with the social network and pursue paths to resilience. The cyclic rhythm of life—connection-disconnection-reconnection—can be found in nature over and over, from sunrise to sunset to sunrise, from the seasons to mating cycles, from birth to death to new birth, and more recently discovered mechanisms of DNA replication, the very essence of life everlasting. No matter where the curious mind explores nature, from microcosm to macrocosm, the rhythm of rebirth can be found, mirroring the hope of the impermanent anxious self that it may attain something of the immortal—Phoenix-like.

The Phoenix Myth represents the rebirth of the impermanent self, the reconnection of dysfunctional social systems and the reorganization of crumbling cultures. Myths are stories derived from observations of nature that aim to explain the workings of the world and man's place in it—they are the glue, the structure that reconnects human beings after chaos disconnects them from each other. Myths are the stories that are told from generation to generation about the origins and resilience of the group—what can be called the Phoenix Phenomenon. As one cultural structure disorganizes as a result of stress from within or without the system, a new reorganization is mobilized that will allow for fragments of the previous structure to reconnect. Art plays a critical role in facilitating cultural transitions during times of stress and adversity (see below "Cave Paintings . . .").

CLAUDE LÉVI-STRAUSS'S THEORY: MYTH AND MEANING

Structural anthropologist Claude Lévi-Strauss made critical contributions to understanding human connectedness—social structures organize in systematic ways that can be understood. Preceding the recent advances in neuroscience, he proposed a system of "binary operations" (*yes/no, on/off, good/bad, beginning/end*, etc) that human beings use to make distinctions about the world, and that these distinctions provide the roots to meaningful narratives that form the securing fabric of culture.

It has now been established that the brain uses this kind of computational system as a basis for all neural functioning: neurons are either "on" or "off." Of course, the trillions of on/off circuits in the brain make for incredible complexity and endless possibilities in interpretation of sensory experience. The brain's left and right hemispheres coordinate the integration of concrete and abstract ideation, and the reverberating connections between the brain and body generate the perception of emotions as derived from fluctuating body states (fear, anger, confusion, etc).

Lévi-Strauss proposed that prewriting cultures are much more attuned to the concrete sensory world of the here and now as compared to cultures with writing that fix abstract narratives in writing for contemplation. In both types of culture, human beings need to find meaningful explanations for events in life. Verbally based cultures generate meaningful stories that explain the holistic totality of the environment and man's place in it so as to soothe the ever-present anxiety of impermanence. Cultures with writing compartmentalize information and need to construct models of explanation for the mechanisms of how the world is ordered, whether they are religious mechanisms or scientific ones. All cultures need to generate meaningful stories of their origin and destiny to keep the stresses of living adequately regulated. Before writing systems developed, myths were passed on by gesture, enacting, storytelling, and images—a picture is worth a thousand words.

Present-day artists face a need from an anxious world to generate meaningful artworks that will "make sense" of the perceived chaos of today's shrinking world. The challenge to artists is to speak to an emerging global culture with art that leaps beyond art history and into a new world of what art can be. Each new artwork contains its own unique meaning. Like a myth, art today is being created out of the stuff of science, religion, mass media, economics, globalization, and many more domains of knowledge—hence the name "Consilient Art."

CAVE PAINTINGS AND DIAMOND SKULLS

The human creative response to chaos has become concretized in the images and objects created by artists. From the caves of Lascaux to Damien Hirst's 2007 diamond-encrusted platinum skull, artists have been creating images and objects charged with meaning beyond their explicit content. The personal and social meaning of art inspires reflection, the critical mental operation for resilience to adversity whether it is individual or cultural.

The cultural value of art outlives the artists, providing timeless possibilities for interpretation by new generations. The Paleolithic cave paintings continue to inspire awe, and speculation as to their meaning will never be considered complete. Theories range from the idea that the paintings functioned as "sympathetic magic" to aid in the hunt and avert starvation—soothing the fear of death, to the emergence of the division of labor allowing talented artists to devote their time to art-making, to an abundance of food that provided more leisure time for the community and a hierarchy of social status—art as a sign of social status and power. Art-making requires a high order of cognitive-social functioning to produce images and objects that invite reflection, inspire awe and wonderment, and can provide individual and social soothing for the anxiety of impermanence.

Reactions to Hirst's skull, titled *For the Love of God,* ranged from "tasteless" to "reflective of the celebrity obsessed culture." When asked why he made such an object, he said, "Because I could." The title of the piece comes from Hirst's mother, who asked her son, "For the love of God, what are you going to do next?"

Created in a culture of global anxiety, seemingly addicted to greed-driven excesses and exhibitionism (from the cartoonlike personalities of Flavor Flav, to self-destructive icons like Michael Jackson and media-crafted moguls like Donald Trump), Hirst's excesses do something that only art does—inspire reflection on multiple levels of meaning—something desperately needed in contemporary society.

The provocative nature of the piece triggers extreme emotional reactions from praise to disgust—the artist sold shares in it to keep part ownership for supervising the public exhibition of the work in museums around the world. The artwork and artist have taken on multiple meanings that cannot be reduced by a simple critical explanation—everything about this work interweaves with the culture: obsession with accumulation of wealth and fame, fear of death, breaking taboos (using the original teeth from the skull of a thirty-five-year-old man who lived over two hundred years ago used to cast the platinum skull), overt embrace of "bad taste" as compared to previous values, art-as-investment commodity, art-as-blockbuster exhibition, excessive consumption, gluttonous decadence of culture, misuse of financial resources, "more is better" attitude, etc. Because the artwork is so provocative, it forces reflection on our culture and may inspire the collective need to make changes in values.

The price of artwork has become more newsworthy than the artwork itself—the fact that a world financial collapse in 2009 was driven by excessive greed and poor reality-testing by the most senior members of the world financial community adds an element of timeliness to the meaning of this artwork—the diamond skull as a marker of the death of greed-driven world culture that is on a suicidal course of overconsumption of limited resources and intoxicating the atmosphere that is killing the planet, our life-support system. One hopes that the planned blockbuster exhibition of Hirst's skull carries with it the message of awakening to these serious issues—thus fulfilling the potential role of art as an agent for cultural resilience—a "Phoenix Phenomenon."

Chapter 4.0

CONSILIENT ART: TOWARD GLOBAL HUMANISM

The period of transition from one system to another is a period of great struggle, of great uncertainty, and of great questioning about the structures of knowledge . . . The tasks before us are exceptionally difficult. But they offer us, individually and collectively, the possibility of creation, or at least of contributing to the creation of something that might fulfill better our collective possibilities.

—**Immanuel Wallerstein**

4.1 Q: WHAT COMES AFTER THE END OF ART?

A: CONSILIENT ART!

"The End of Art" discourse emerged in art criticism at the end of the twentieth century.[24] Marcel Duchamp, Barnett Newman, and Andy Warhol were seen as artists whose works and theories were critical to the "end of art" era. The idea that "aesthetic art history" in Western culture emerged during the early Renaissance out of an ancient history of "sacred artisanship" for religious purposes (icon paintings, mosaics, sculpture, temple architecture, etc.) seems to be coming to an end. Art had functioned for centuries as an object of sacred reverie and personal

---

[24] D. Kuspit, *The End of Art* (Stony Brook, NY: Cambridge. 2004).

aesthetic and spiritual reflection. (Eastern cultures made little distinction between aesthetic, sacred, and decorative arts, and saw all art forms as a continuous expression of human creative ingenuity.)

As art in the twentieth century became more anti-aesthetic, transforming into an investment commodity or entertainment medium, artists and critics voiced concerns about a breakdown in the boundary between aesthetics (high art) and commerce (commercial art). Some feared the loss of aesthetic meaning in art.[25]

Simultaneous with these "end of art" discussions, two great scholars published what might be called their manifestos or summation of their life's work: evolutionary biologist E. O. Wilson's *Consilience: The Unity of Knowledge* (1999) made an urgent plea for bridging all domains of knowledge with fact or fact-based theories as had occurred in the sciences, and Immanuel Wallerstein's *World Systems Analysis* (2004), calls for "unidisciplinary" integration of the historical-social-economic sciences. Both observed that the emerging contemporary global culture had its origins in the Renaissance and had flourished for four hundred years but was entering a process of radical transformation that required a new, unified knowledge to avoid total chaos and disorganization.

Wilson focuses on the fragile biosphere—the planet as a single living system—being threatened by the gluttonous human consumption and destruction of the life-support system. Wallerstein parallels this concern with a focus on the capitalist economic system's "endless accumulation of capital" exhausting limited planetary resources. Both Wilson and Wallerstein, septuagenarians, had lived through World War II and had seen the "bang" that could destroy the world. They are now devoted to calling the alarm for new ways to think about how we can creatively adapt to the inevitable global human society—to avoid the "whimper" described by T. S. Eliot—"This is the way the world ends, not with a bang but a whimper."[26]

The world financial crisis and the ever-present news about global terrorism amplify the theories of a world-system-in-transition put forward by the experts in art history, world-systems analysis, and evolutionary biology. Global anxiety and insecurity have never been felt so universally—no one is immune to the potential catastrophes that are possible. However, these global adversities are also kindling the human Art

---

[25] A. Danto, *After the End of Art* (Princeton, NJ: Princeton, 1998).

[26] T. S. Eliot, *The Hollow Men* (1925).

Imperative with new aesthetic experiments and explorations—the result of these experiments is new art that bridges domains of knowledge—what I will call Consilient Art. Art-making has always been critical to cultural survival as old social systems are displaced by the evolution of new systems.

4.2 ART AS AGENT FOR TRANSCULTURAL RESILIENCE

On September 11, 2001 terrorist destroyed more than 2 buildings and the human beings there in. A new world stage arose from the ashes. The stale conflicts between communism and capitalism that dominated world culture in the last half of the twentieth century and posed an ever-present global anxiety of nuclear holocaust were all but gone. Globalism seemed like a benign inevitability, and most Americans could not find Kuwait or Afghanistan on a map, much less understand the difference between hummus, the food, and Hamas, the Palestinian Islamic resistance movement. The terrorist attack that made the crumbling of the World Trade Center the most looked-at event on television for months exceeded the greatest hopes of any terrorist—it created a lasting mental image of chaos, fear, and death in the minds of everyone, much the way the mushroom cloud over Hiroshima had done. A new world conflict dominated the stage of history—religious fundamentalism vs. secular humanism—plunging humanity into a global version of the age-old tribal battles between neighboring cultures. Marshall McLuhan's idea of a "global village" has become a reality, with neighboring global villages waging interminable guerilla warfare against one another.

Art has often played a role in cross-cultural connections due to the novelty and exotic nature of foreign cultural artifacts. Art has been made as offerings of peace and bids for trade agreements, dating back to Marco Polo and beyond. Art has been pilfered by conquering invaders from the Spaniards in Mexico to the British in Egypt to the Nazis in France. In the present world, artists have the potential to create images and objects that displace the images of fear created by global conflict from mass consciousness and replace them with images of creative adaptation and resilience. The emergence of new art forms, new media, new venues for exhibition, and new collaborations between artists and experts from science and technology may empower art to play a significant role in world-stress management.

New Aesthetic Typologies: Artists as Brain-Structure Equivalents

A new typology of art can be derived from the mechanisms of certain brain structures and functions to understand the parallels that art may play in facilitating cultural resilience. An analogy to certain brain-structure functions can be made based on the content of the artist's work. By understanding how the brain coordinates neural networks for adaptive functioning, art can be categorized from the "bottom up"—a term used to designate information sent up to the brain from the body's sensory systems—rather than by traditional art-historical "top-down" typologies—from the brain's reflective prefrontal cortex to the lower brain structures and the body.

The mammalian brain evolved specialized structures to function in coordination with each other for optimal adaptive resilience to the adversities faced by the organism: cognition, emotion, motivation, and regulation of bodily functions (fig. 17). Human social systems evolved a division of labor to organize complex group behaviors that empowered survival, security, and the pursuit of happiness. From this list of brain structures and functions, it is clear that artists utilize their entire brain and its interconnections in the art-making process. Because of it meaningful ambiguities, art cannot be understood by a reductionistic model of brain function:

CEREBRUM

Frontal Lobe

Behavior
Abstract thought processes
Problem solving
Attention
Creative thought
Some emotion
Intellect
Reflection
Judgment
Initiative
Inhibition

Coordination of movements
Generalized and mass movements
Some eye movements
Sense of smell
Muscle movements
Skilled movements
Some motor skills
Physical reaction
Libido (sexual urges)

OCCIPITAL LOBE

Vision
Reading

PARIETAL LOBE

Sense of touch (tactile sensation)
Appreciation of form through touch (stereognosis)
Response to internal stimuli (proprioception)
Sensory combination and comprehension
Some language and reading functions
Some visual functions

TEMPORAL LOBE
(LIMBIC STRUCTURES: AMYGDALA, HIPPOCAMPUS, ETC.)

Auditory memories
Some hearing
Visual memories
Some vision pathways
Other memory
Music
Fear
Some language
Some speech
Some behavior and emotions
Sense of identity

RIGHT HEMISPHERE
(THE REPRESENTATIONAL HEMISPHERE)

Control of the left side of the body
Temporal and spatial relationships
Analyzing nonverbal information
Communicating emotion

LEFT HEMISPHERE
(THE CATEGORICAL HEMISPHERE)

The left hemisphere controls the right side of the body
Production and understanding of language

CORPUS CALLOSUM

Communication between the left and right side of the brain

HYPOTHALAMUS

Moods and motivation

THE CEREBELLUM

Balance
Posture
Cardiac, respiratory, and vasomotor centers

THE BRAIN STEM

Motor and sensory pathway to body and face
Vital centers: cardiac, respiratory, vasomotor

ARTISTS AS CULTURAL-BRAIN STRUCTURES

JEAN AUGUSTE DOMINIQUE INGRES: ARTIST AS PREFRONTAL CORTEX AND HIPPOCAMPUS

Ingres' famous portrait of Napoléon (fig. 21) functions as an archival record of psycho-sociopolitical history charged with all of Bonaparte's grandiosity—his mastery of realistic painting transports us instantaneously into the presence of Napoléon himself, communicating a story of the man, his personality, and his times far beyond an essay about him. Ingres functions in his culture parallel to his brain's coordination of frontal lobe and hippocampus—the structures involved in the selection and formation of new memories. By archiving emotion, cognition, and motivation in a single picture on a grand scale, Ingres utilizes the brain's capacity for creating a complex visual memory for the ages. Artists-as-archivists perform a critical role in the consolidation of complex meanings for the reflection of future generations.

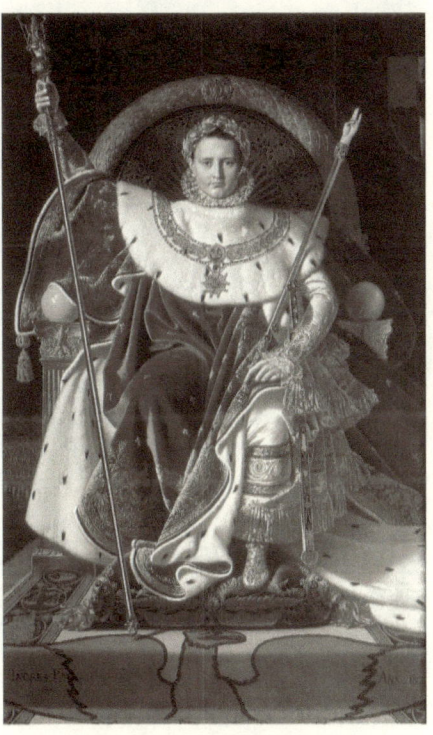

Fig 21 Ingres Napoleon

HANNAH HÖCH: ARTIST AS AMYGDALA

Another example of Artist as brain structure is Hannah Höch: her Dada collages function as a PFC-amygdala (the structure that triggers the stress response) feedback loop, alerting consciousness to imminent calamity. Höch's 1920 pasted-paper collage, titled *Schnitt mit dem Küchenmesser Dada durch die letzte Weimarer Bierbauchkulturepoche* ("Cut with the Dada Kitchen Knife through the Last Weimar Beer-belly

Cultural Epoch in Germany") (fig. 22), pioneers photomontage crafted from images and text printed in German newspapers.

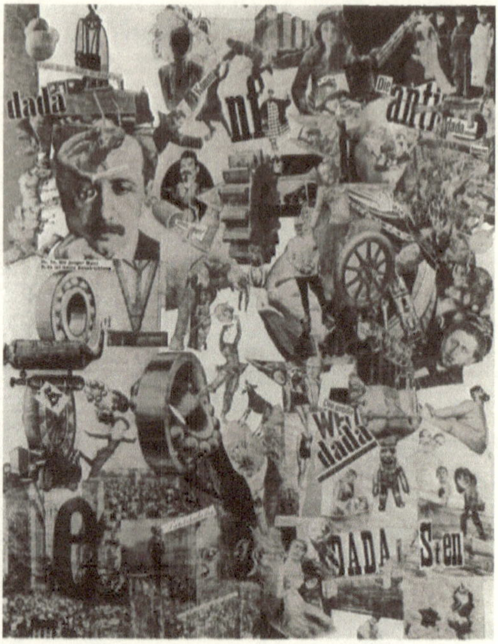

Fig 22 Hanna Hoch

Cut with the Dada Kitchen Knife expresses the Dada movement's protest against European "high culture" by recontextualizing found media images and framing them as farcical. Höch's collage is iconoclastic and anti-art-establishment—depicting one male figure wearing a skirt; another has a donkey resting on the top of his head; a third female figure appears to be wearing a hat intended for a male that is much too large for her.

CIMABUE, GIOTTO, AND THE HUMANIFICATION OF CHRIST: ARTISTS AS ORBITOFRONTAL CORTEX

A third example of artist-as-brain-structure is the work of Cimabue and Giotto (mentored by Cimabue), whose naturalistic painting of Christ transformed the iconic Byzantine style of idealized images of the Messiah that were more cartoonish.

Like the orbitofrontal cortex that makes choices and learns to choose adaptive behaviors over repeating maladaptive habits, Cimabue and Giotto began a new visual tradition that could have resilient power for individuals and culture. By painting the crucified man on the cross to look like a suffering person one might meet—the humanification of Christ—Cimabue (1240-1302) and Giotto (1267-1337) brought Christ closer to everyday reality in a time decimated by the bubonic plague (1328-1350) (fig. 23).

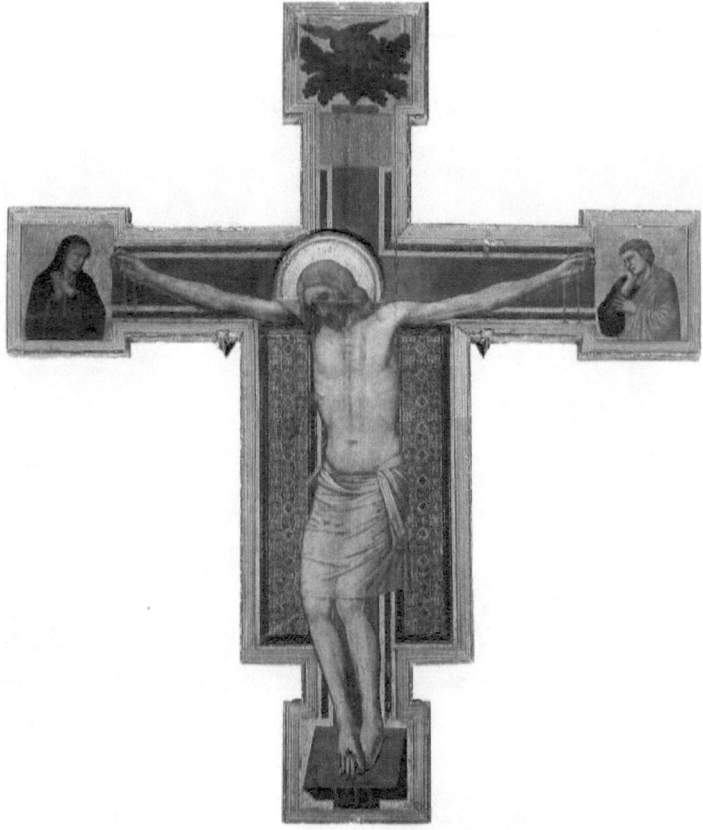

Fig 23 Giotto Crucifixion

The Christian culture must have been in serious doubt about their God unleashing the plague. By rendering Christ as an ordinary man, Cimabue mirrored the human suffering of life in the 1300s with the suffering of Christ—this revolutionary imagery implicitly communicated that ordinary people were one with Christ. It must have had far-reaching powerful soothing effects on anyone who saw it. The collective need to bring Christ closer to home may have contributed to why this style was so widely supported by the church and accepted by the culture. This naturalistic painting style functioned as cultural resilience to the devastating effects of plague and starvation and shaped Western art history for the next seven hundred years.

Giorgio Vasari (1511-1574), painter, architect, and biographer of artists, wrote that Cimabue's art "astonished the world in those days, painting having been so long in such darkness, and to myself, who saw it in the year 1563, it appeared most beautiful, and I marveled how Cimabue could have had such light in the midst of such heavy gloom."

RECOVERING THE SACRED IN ART: MICHAEL TRACY

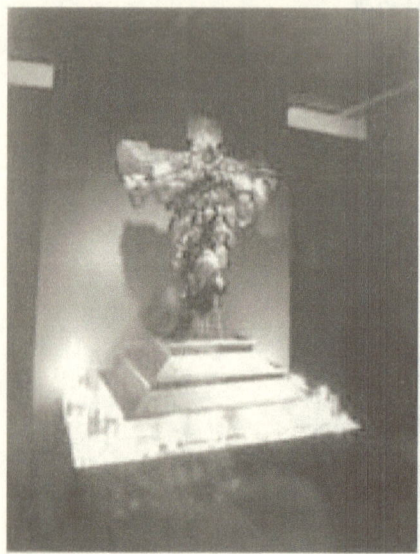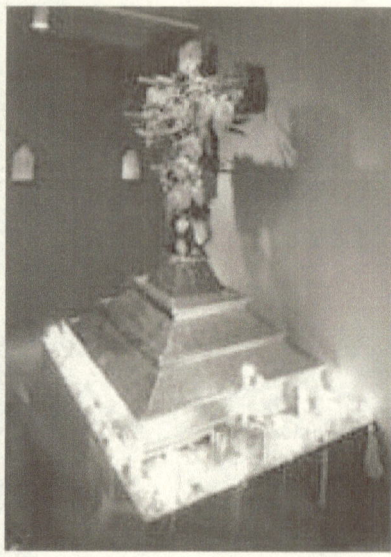

Fig 24, A, B C, Michael Tracy, *Cruz de la Paz Sagrada, 1980*

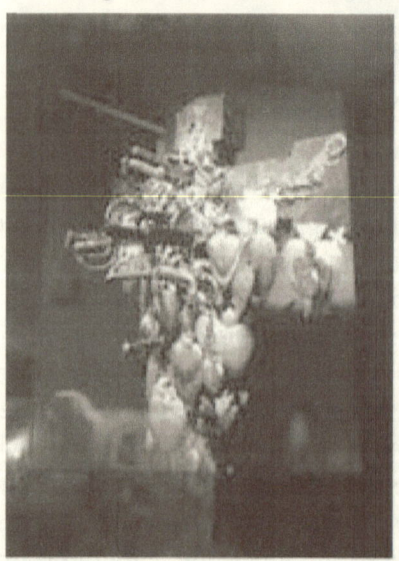

Of all the stresses that face individuals and cultures, the most heinous is the human capacity to hurt, brutalize and torture each other. From parental child abuse to genocide, we must perpetually struggle with this trait for inflicting pain on each other.

The theory that Cro-Magnon's emergent cognitive powers for abstraction and future planning empowered his primordial survival instincts to become the most creative and destructive force on the planet casts a long shadow across human history. It is likely that Cro-Magnon exterminated Neanderthal, his chief rival for survival during the same epoch that he created the magnificent cave frescoes of Altamira, Chauvet and Lascaux. This cognitive power is a double edged sword that threatens the human species in the present day conflicts between between militant fundamentalist religious groups and the reigning powers of global secular humanism. Violent, destructive actions against each other prevail on a global scale as never before. The power to destroy ourselves and to create art spring from the same evolutionary brain capacity. It is imperative for our survival that we chose creativity over our survival driven destructive urges—we must continue to make new art for reflection on our selves, our cultures, our behaviors and values.

Michael Tracy has devoted his life to recovering the sacred meaning of art—placing suffering humankind in role of the sacred. By challenging the unstoppable secular commodification of art embodied in the works of Andy Warhol and others—art as a business product—Tracy has created a singular body of work that aims to elevate human beings to a state of sacredness. Emerging from the abstract expressionist movement, his efforts to recover the sacred in art focus on human suffering. Utilizing devalued iconic forms from Christian art history Tracy reframes the socio-political domain of human suffering—man against man—placing human life as worthy of sacred respect. Tracy's work invites reflection on a radical reframing of religious dogma that elevates human life to the realm of the sacred. Never in human history have we had a greater need for art to function as a mirror of our capacity for mankind's inhumanity to mankind—Michael Tracy's work is such an imperative mirror.

4.3 CONSILIENT ART: TOWARD GLOBAL HUMANISM

The term *Consilient Art* identifies artworks that trigger reflection by integrating different domains of knowledge that charge the art with meaning beyond the explicit content. The artists often aim to inspire mystery, awe, wonderment, and other intense emotions, be they negative or positive.

The Art Imperative hypothesis sees this need to create new, startling images as a virtual laboratory for the culture to experiment with change, the inevitable stresses that can undo cultural traditions and stability. People and cultures that suppress art and creativity have a very good reason to do so: creativity requires use of the reflective functions of the brain, and reflection threatens the status quo, tradition, and the religious dogma of the person or culture.

By limiting or suppressing creative expression, such cultures enter a period of stagnation and tend to view the changes in other cultures of the "outside" world as hostile or threatening. It is clear that the global systems we live in now are in chaos, conflict, and transition. The greed-driven financial collapse of the global economy has kindled clashes between stagnant fundamentalist cultures from all religions—they are fighting for life, and the unstoppable advances in secular culture's technology and communication are shrinking the world into a "global village" that is in violent, destructive turmoil. This global "end-of-cycle" phenomenon is kindling sections of the art community to ignite their creative fires that will consume and transform the art of the old cultures, delivering possibilities for the rising of a new cultural Phoenix.

Consilient Art has been created throughout history from the art-engineering-religion integration of the elaborately decorated Egyptian tombs to Damien Hirst's cross-sectioned animals in formaldehyde. Hirst emerged in the 1990s and draws widely from the sciences and religious iconography often carrying an explicit intention of shock and awe. With the passage of time, the shock value of Hirst's art has diminished, but the soaring prices of his art continue to awe the media and the public. Consilient Art breaks boundaries and reframes what art can be—Hirst's shark in formaldehyde is an expected exhibit at a museum of natural history, but recontextualized as a work of art, this same shark takes on multiple meanings that inspire reflection on a personal and cultural level.

Many new artists are emerging in the wake of Hirst's science-religion-art fusions, incorporating artifacts, ideas, and methods from the sciences—reciprocally, science is becoming increasingly involved in the study of creativity in all the arts. Books about the neuroscience of art, music, architecture, dance, ethics, gambling, etc., are being written by researches all over the world. For the first time, science can explore creativity with tools to see the brain during its creative periods. Likewise, these tools and this information inspire artists to explore new media and materials in expressing contemporary issues for human culture.

CONSILIENT ARTISTS: PHOENIX RISING

In 2009, a London exhibition of new art, The Age of the Marvelous, curated and produced by Joe LaPlaca and All Visual Arts, marked a historic moment in the twenty-first century—Consilient Art has arrived in the form a cultural art movement.

Like the mythical Phoenix bird that rises from the ashes of its own death, this new Consilient Art gathering in The Age of the Marvelous is charged with the mysterious power to kindle new life from a dying generation and culture. Seamlessly fusing images and objects from art, science, and religion, this exhibition comes shortly after the world financial crisis in 2008 and Damien Hirst's diamond-encrusted skull (2007) that seems to mark the death of greed-driven culture. These new artworks startle the viewer's attention, trigger the imagination, and inspire wonderment. New art that simply shocks but fails to hold the viewer's attention will not imprint a potent memory in the brain to contemplate.

The "idea translations"[27] created by these artists' reflective brain functions explore, experiment, and take risks in the creation of new art— the viewer is implicitly invited to bring an open mind to the encounter, allowing the art to trigger his/her brain's contemplative function and make their own idea translations. This interaction between the artist, the art, curator, and the viewer ripples through culture in private conversations, news media, and certainly among other artists. The emotional and cognitive content of this new art carries an implicit message: a complex transformation in art is happening, here, now, and in London! The vector for kindling cultural resilience to adversity lies in the conscious and nonconscious emotions and motivations triggered

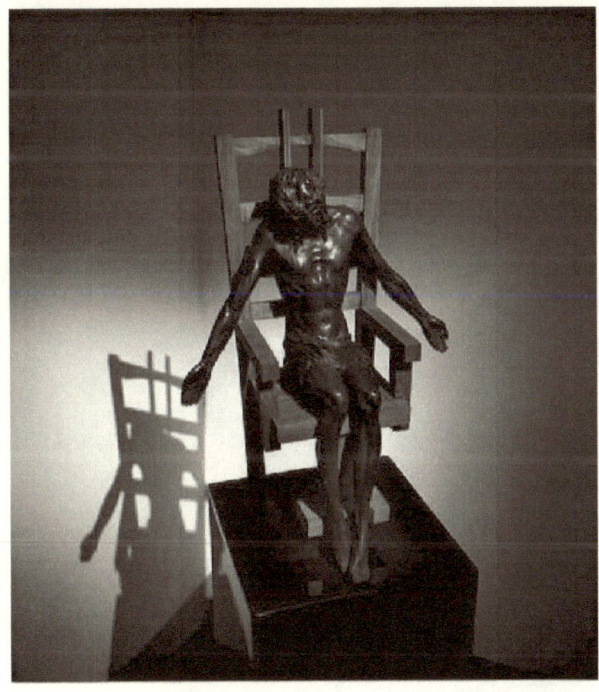

Fig 25 Paul Fryer *Pieta*

---

27   D. Edwards, *Artscience: Creativity in the Post-Google Generation* (Cambridge: Harvard University Press, 2008).

by the art in the viewer's memory. A child witnessing this extraordinary show will no doubt remember it for the rest of his/her life.

It is likely that some of these works will offend and confuse some viewers: Paul Fryer's *Black Pieta,* a black Christ killed in an electric chair (fig. 24), and his life-size crucified primate staged in the central apse of the Holy Trinity Church; and Polly Morgan's *Pyric Victors,* a small infant-sized coffin, burnt black and disintegrating from one end while a swarm of tiny gray and black taxidermied chicks emerges Phoenix-like from the other end. Joe LaPlaca chose to exhibit The Age of the Marvelous in a historic church designed by John Soane and has arranged the artworks so that the relationship to the architecture and to the other artworks amplifies the overall exhibition beyond the power of any individual piece. This is an implicit exhibition about relationships in time and across time. The works themselves offer unique reconfiguring of just what art can be—each artist seemingly on a personal mission to reinvent art in personal and historic terms.

ART AND ARCHITECTURE: A DIALOGUE BETWEEN ALYSON SHOTZ AND SIR JOHN SOANE

The first art relationship encountered is the external dialogue between Alyson Shotz and Sir John Soane. Shotz's *Helix* (fig. 25), a towering sculpture of forty-one pairs of spiraling squares stacked with

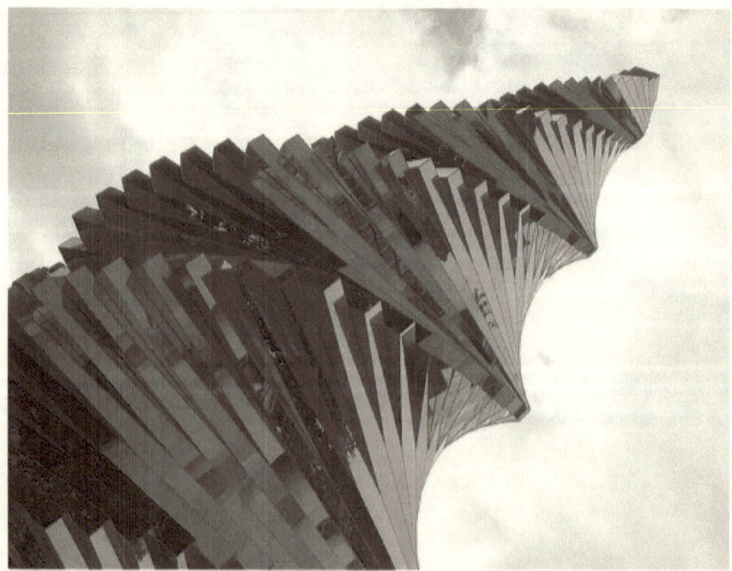

Fig 26 Alison Shotz *Helix*

architectonic precision and covered in dichroic film that refracts light and reflects different colors based on the viewer's position, invites the viewer to circumambulate the work like a Buddhist praying as they walk around a sacred shrine.

Helix, a science-art-architectural rainbowlike monument, stands at the front right of Soane's monochromatic neoclassical church, with its quiet dignity, soothing simplicity, and perfect balance. Off in the London skyline rises the Telecom Tower, a beacon for satellite communication, that aligns with the church and the sculpture in a straight line.

Coincidence? As clearly disparate in time as these structures are, they harmonize in space with a voice that beckons, "Please come in."

During the evening, the spotlight on the mirrored surface of *Helix* reflects an abstract pattern of light back onto the outside walls of Holy Trinity Church—the opposite of shadow. Viewing Shotz's work in a white box gallery would surely be an encounter with the beautiful, but its power to command attention would not feel quite so strong as it does here at One Marylebone Road.

THE GRAND CHAPEL AND ADJOINING CHAMBERS: CONJURING SPIRITS OF THE PAST

Entrance to the foyer, where the devoted congregation would gather every

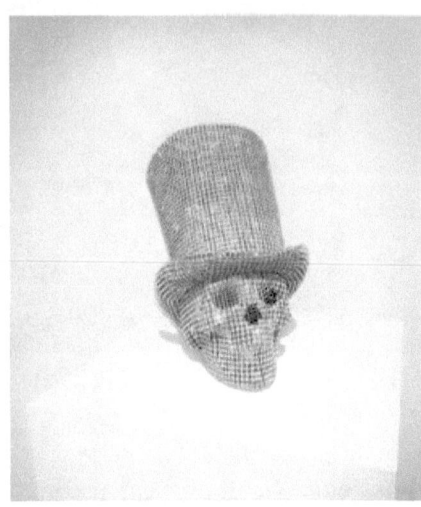

Fig 27 Wolfe von Lenkiewics
St Eustice

Fig 28 Nicola Bolla *Vanitas Skull with Tube Hat*

Sunday, is greeted by two sculptures in the adjoining rooms to the left and right: Wolfe von Lenkiewicz's life-size bronze stag's head with a "terrorist" airplane stuck in its head, *St. Eustace* (fig. 26), and Nicola Bolla's life-size crystal human skull with a top hat, *Vanitas skull with tube hat* (fig. 27). They seem to face each other, flanking the viewer like guardians of the exhibition—the post-traumatic but saintly animal world to the left, and a glistening jester of death to the right. Each sculpture sits Sphinx-like at the stairway up and down to the other levels of the exhibition.

In a dimly lit chamber, a contemplative gray palette dominates and invites us into the reflective world of water—a gathering of five works. Ben Tyers's *Breathe* (fig. 28) looks like an egg-shaped fishbowl half-full of water, but it is in fact a device that breathes. Like a yogic metronome, each inspiration sound one hears is accompanied by the water level falling as the air is sucked into the empty portion of the vessel, and with each soft expiration sound the water level rises like a human pulmonary diaphragm pushing out the air through the "airway" at the top of the

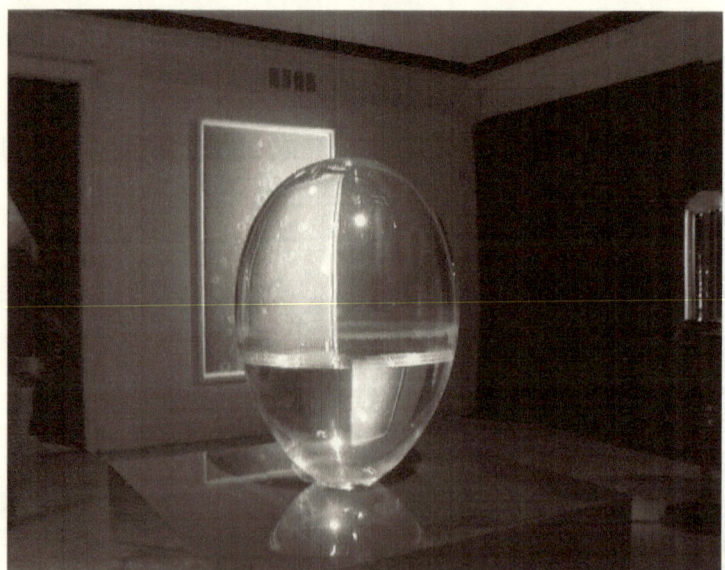

Fig 29 Ben Tyers *Breathe*

glass egg. Standing near it for even a few seconds triggers one's own breathing to synchronize with this "guided meditation apparatus."

Flanking this "living machine" to the left sits Alyson Shotz's *Ice Network,* a delicate and colorless matrix of glass that looks like icicles, and to the right Alastair Mackie's paradoxical and black-humored *Metamorphosis,* a mirrored 1930s taxidermy display ell jar, reflects the viewer rather than presenting a stuffed animal.

Mounted on the wall opposite these three sculptures and flanking the doorway into the main exhibition chapel are two of Adam Fuss's unique gelatin silver print photograms, *Untitled.* This gray diptych looks like rippling circles of water droplets just after striking the surface of water. The photograms freeze the movement of rippling water and offer a counterpoint with the moving water in *Breathe* and the frozen water of *Ice Network.* The cylindrical mirrored surface of the bell jar reflects the entire room with its "states-of-water" art and the viewers like the surface of a perfectly still pool. The presentation of these watery artworks invites the viewer into a visual meditation on "states of consciousness."

The grand chapel of Holy Trinity Church, a masterpiece of proportion and light, holds twelve works by seven artists—the new congregation in this Consilient Chapel—spaced in an almost bilaterally symmetrical manner, reminiscent of the pews that once seated the faithful. Hanging low in the center of the nave of the chapel is Polly Morgan's *Departure,* a gigantic brass "chandelier" inspired by a Victorian fantasy for a flying machine where the "pilot" would sit in a cage and the apparatus would be carried into the sky by birds harnessed and controlled.

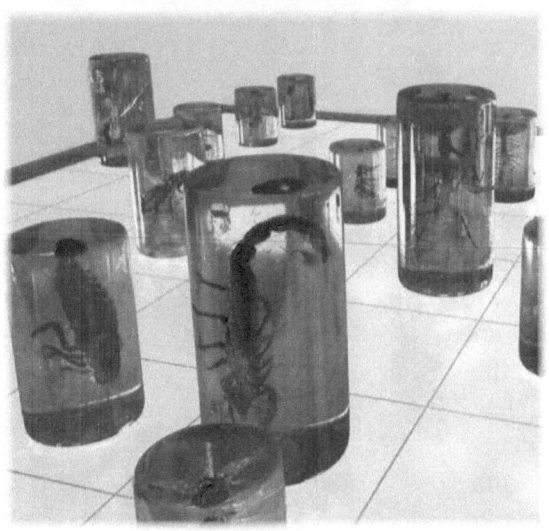

This awesome spectacle is surrounded by nine works in the right and left aisles of the nave. Alastair Mackie's two works begin the parade: to the right is his *Amorphous Organic* (fig. 29), the auto-illuminated chess set with pieces made from amber-entombed insects. To the left is a perfect sphere of

Fig 30 Alistair Mackie *Amorphous Organic*

mouse skulls glued together, creating a gravity-defying challenge to the imagination.

Further down the left aisle, we find Kate MccGwire's duo of beautifully crafted bio-form sculptures made of feathers mounted on polystyrene, *Urge* and *Wrest* (fig. 30). Across to the right and just after MccGwire, we find a pair of works by Hilary Berseth: *Programmed Hive #9,* created by bees that colonize an artist-designed armature and configure it with honeycomb to suit their needs, and *Untitled 2,* a matrix of electroplated copper wire that was "grown" in a bath of copper-saturated solution.

Completing the parade down the left aisle is Alastair Mackie's *House,* a dollhouse structure made from three hundred pulverized wasp and hornets' nests (homes). Across the nave is Hugo Wilson's *Parabiosis,* a polyurethane, resin-glass, and wood sculpture inspired by the anatomical dissection of two interconnecting cardiovascular systems of Siamese twins, an utterly codependent anomaly of nature, and Paul Fryer's sculpture, *Telstar,* an exact copy of the first communication satellite ever launched into space rendered in marquetry wood inlays by master craftsmen.

The presentation of stuffed birds, the amber-cast insects, and the mouse skulls requires no imagination to associate with death, but the re-presentation of these artifacts of life transformed by skilled hands and aesthetic mastery into chess pieces, spherical sculpture, and a flying machine is truly wondrous to behold. The abstract feather sculptures, cardiovascular twins, honeycomb castle, house of wasp and hornets' nests, wooden marquetry-crafted satellite bring us back to an awareness of the cyclic-impermanence of our bodies, the warmth of our homes, and just how important handmade objects are for our reverie and contemplation. The chromatic coordination of these organic

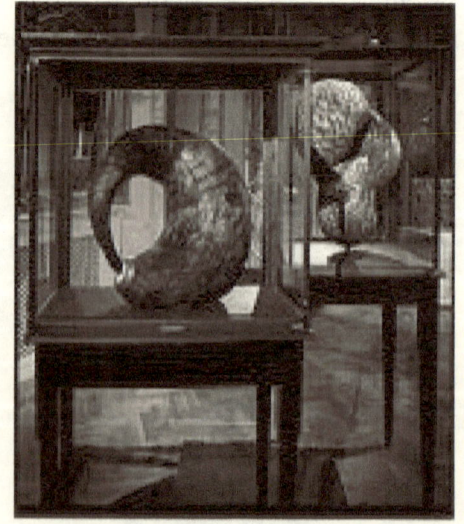

Fig 31 Kate MccGuire *Urge and Rest*

objects and the soft spotlighting create a visual choir where the voices of the congregation once sang the praises to their Lord. Now the lighting casts dramatic shadows of the artworks on the walls of the nave, reminding us that we live in the shadow of lost generations, forever seeking and creating new light. Like the individual voices of the long-lost choir, the singularity of these artworks amplify each other in a harmonic transformation of materials and synchronistic creation of new meaning. Consilient art does so much more than shock or startle—that is only the beginning—the complex ideas and skill at presenting different domains of knowledge in beautiful and challenging forms inspire wonder and stimulate the imagination.

In the elevated presbytery, where the elders of the church would sit during services, Keith Tyson's *Contemporary Grotesque: Mastering* (fig. 31) confronts us with a life-size Japanese geisha sitting atop a lumbering walrus, a witty allusion to man's ritualistic attempts to dominate nature. Cast in polycarbonate with a graphite patina, Tyson focuses on carbon, the substrate for all organic molecules and for his art. The choice of

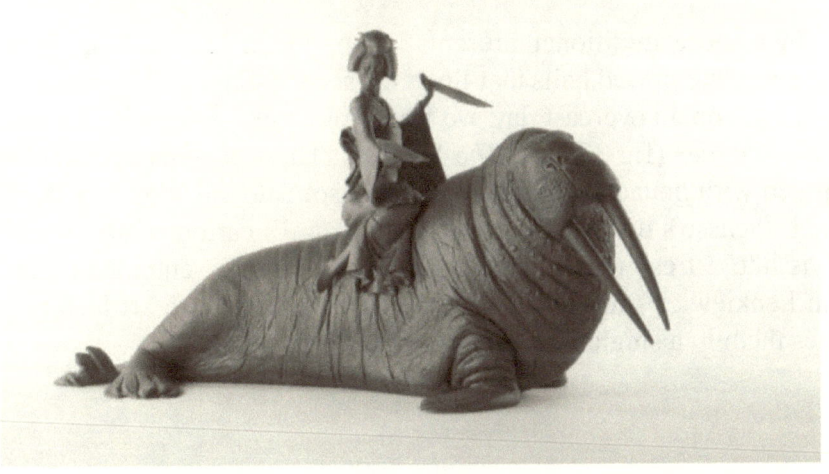

Fig 32 Keith Tyson *Contemporary Grotesque*

materials conveys the artist's curiosity about the "molecular sociology" of human beings—that is how events at the molecular level might effect our social life and vice versa.

Rising in the apse, the most sacred place in the church, we are confronted with Paul Fryer's crucifix, *A Privilege of Domain,* a life-size sculpture of a crucified waxwork protohuman primate (the missing link in evolution?) that looks chillingly real. Clearly designed to provoke a powerful emotional response of compassion and outrage, the work invites contemplation of one of the most powerful global debates between science and religion: Darwin's theory of evolution versus the fundamentalist faith in "intelligent design."

The massive sculptures of Tyson and Fryer juxtapose historical figures and animals in a topsy-turvy fashion, forcing new connections in the viewer's expectation of art. Looked at as a duo, these works reverberate the confusion in our secular world about personal values, ethics, our relationship to other living creatures and to each other, and the role of spirituality in our daily lives. The sermon that seems to be delivered by these two sculptures from these holy spaces in the chapel seems to be aimed at contemporary culture itself. It might be, "Reflect, contemplate, reconnect with your essence, and deliberate before making choices about what direction to take in the future."

RECONSOLIDATION OF ICONIC MEMORY: THE MAGICAL WORKS OF WOLFE VON LENKIEWICZ

In a more traditional presentation provided by Sir John Soane's glistening windowed halls that flood the second-floor gallery space with light even on an overcast day, we find Wolfe von Lenkiewicz's diptych, *Ace of Spades* (fig 32) and *The Smokers*. These two massive canvases painted with gouache and charcoal reconsolidate the "dots" of Damien Hirst, Picasso's distorted portraits, Murakami's cartoons, and Warhol's Elvis into a fresh new image. Like Mackie's amber-entombed insects, von Lenkiewicz's masterful hand captures heroes from art history and fixes them in a single new surface for reflection.

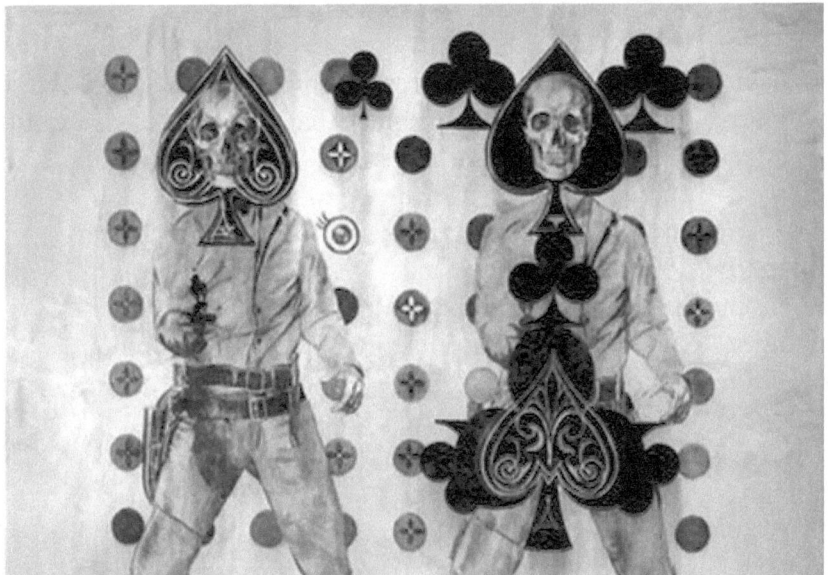

Fig 33 Wolfe von Lenkiewicz *Ace of Spades*

Von Lenkiewicz is engaged in a project of reconsolidating cultural icons both past and present, liberating them from their temporal origins and popular meanings. By synthesizing art icons created by Picasso, Giotto, Warhol, Chagall, da Vinci, Disney, Hirst, and more, Lenkiewicz triggers the memory and startles the imagination.

Human memory was once thought to function like a CD-ROM (read-only memory). No matter how many times you play it back, the memory was unaltered. But memories actually become malleable and changeable during the process of being recalled. Memories in long-term storage only stay stable up until they are remembered. The process of remembering alters stored memories physically so that they become temporarily unstable and need to be "reconsolidated" or made stable once again. By remembering the past, we create a new, slightly altered memory, charged with experiences from the present—keeping the essence of past fresh and alive.

Wolfe von Lenkiewicz's images directly embrace the Art Imperative hypothesis to create meaningful new images from the old—a critical practice for resilience to the stresses of change—and the quintessential embodiment of the Phoenix phenomenon.

The Crypt: A Consilient Laboratory-Studio

No entombed priests or elders can be found here anymore, but the dim lighting sustains a somber atmosphere that once must have prevailed, long after the dead have departed. Now we find new art between the arched columns, kindling a secret intimacy between art, science, and religion in contemplative monastic tranquility. Paul Fryer's notorious *Black Pieta* dominates as the centerpiece of the crypt like a sacred shrine that can only be viewed privately by special arrangement with church elders. The recontextualization of the death of a black-skinned Christ by electric chair rattles all the historical expectations of sacred Christian art and simultaneously answers the oft-heard reflection, "What would Christ do if he were alive today?"

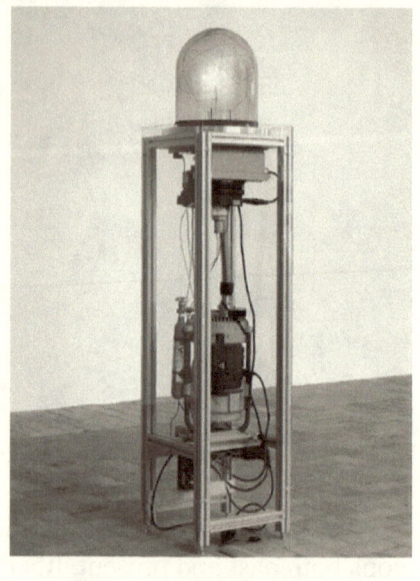

Fig 34 Paul Fryer & Colin Dancer
Evening Star

Behind the *Black Pieta,* Alastair Mackie's *Untitled (+/-)* presents two concrete slabs on the floor, one holding a conical pile of mouse skeletons and mouse fur, and the other supporting a loom with a half-completed shawl or ruglike textile, presumably made from threads spun from the pile of dead mouse debris. *Untitled is* the most concrete example of the show's entire theme: paraphrasing Jasper Johns's famous dictum, "Take something dead, do something to it, then do something else to it.'

Paul Fryer collaborates with many artisans to create his work: wax sculptors, marquetry woodworkers, metalworkers, and especially with physicist-engineer Colin

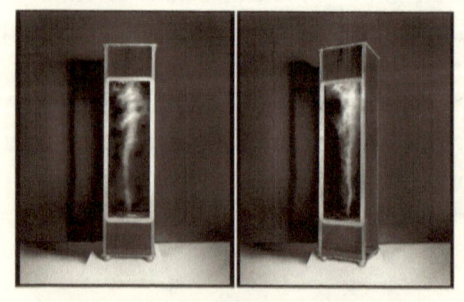

Fig 35 Hugo Wilson *Tornado*

Dancer. Together they create extraordinary science-art objects: *Venus and Mars,* an updated orrery that recreates the movement of the planets in relation to one another, and *Evening Starr* (fig. 29), an apparatus that recreates the physics of the *aurora borealis,* capturing the soft, pulsing light in a bell jar, and *Evening Star* is literally a star-in-a-jar, a gaseous ball of plasma that is ten million degrees centigrade at the center.

Fryer also shows a more classical sculpture dedicated to Marquis de Laplace, one of the greatest scientific minds in history. *For Laplace (Fear),* a perched angel crouching atop a tall spiral staircase with the facial expression of fear, one of the primary emotions, connects scientific curiosity about the unknown with the fear kindled in those that explain everything through blind faith and religious dogma. Science has all but vanquished the life of angels, placing them in the realm of magical belief like Santa Claus.

Hugo Wilson's *Tornado,* a 250 cm. tall black box with a glass front, houses an actual vortex of humidified air that perpetually changes shape despite the programmed controls of the machine.

Maria della Signore's *Quartet (Staying still along its way)* is presented in a pitch-black shrouded room illuminated only by four strobe lights, blinking separately, one in each of four tall glass-box sculptures containing rainlike dropping water. With each flash of one strobe, we can visualize what looks like suspended raindrops, frozen in their fall. The irregular separate flashes stand in a semicircle, allowing the viewer to surround himself with these glimpses of the towers of falling-but-still water, changing position with each pulse of light. Hypnotic, contemplative, and soothing with the sound of falling water, this darkened chamber carries the imagination away in many directions.

Other artists in the crypt have reconsolidated religious themes, images, and objects of time, death, impermanence, and transformation. Martin Sexton presents two famous heads: Socrates and St. John the Baptist. *Diotima teaches Socrates in the ways of love* is a mask of the Greek philosopher constructed from a 4.5-billion-year-old meteorite, wall mounted on a projection of the moon going through all its cyclic phases accompanied by a performance of the oldest surviving example of a complete musical composition. *Levitation of the head of St. John the Baptist* is a small decapitated head of the Christian martyr floating mystically in a sacred reliquary externally painted with a copy of Leonardo's portrait of the young saint.

The great link between art and science is their capacity to make the invisible visible. Both require heightened capacity for attention to detail and enormous patience to arrive at a place of illumination. The imagination of each artist seems to have been captured by the domain of science, inspired by works of past art, reconsolidating imagery and iconography from religion in order to produce works of great beauty and mystery—successfully seducing the viewer into a state of wonderment. The crypt—once a place of death—has been transformed into a living laboratory-studio of Consilient Art—elevating the exhibition of nature's mysterious beauty and cyclic rhythms to the realm of the sacred.

THE AGE OF TRANSFORMATION: REFLECTIONS ON CONSILIENT ART AS AN AGENT FOR CULTURAL RESILIENCE

As the historians tell us, the world we know is in crisis—a challenge to our very survival—undergoing inevitable transformations in the biosphere, the world socio-economic system, and the reframing of aesthetics by a technology-driven collision of cultures. The present day cusp of history finds us all in a world of global anxiety—desperately seeking new ways to solve old problems, be they cultural, political, financial or spiritual. Fear and anxiety trigger freeze-flight-fight responses in individuals and cultures—often mobilizing self-sabotaging destructive actions. Art has the power to spark reflection and kindle creative imagination, like a combination of a scientific laboratory and a contemplative retreat. Artists function as agents of cultural resilience, especially in times of crisis, by creating new works that confront and challenge tradition while inviting curiosity—exploration and creation of new attitudes and possibilities for action is critical for cultures in transition to survive.

The Consilient Art exhibited in The Age of the Marvelous triggers contemplation, erases traditional boundaries between art, science and religion, creating a collective amplification of singular visions into a unified theme of Transformation. Materials, ideas, symbols, images, and techniques are seamlessly woven into an exhibition of subtle and profound power. The yin/yang of this new, consilient art combines 'shock and awe' with 'contemplation and exploration'. Artists reframe the very definition of what art can be—viewers that allow their brains to follow the contemplative trigger offered by art will surely experience the magic of aesthetic awareness. Although darkness prevails as an aesthetic choice in this exhibition, it points toward the fact that after any darkness light will follow.

The success of Consilient Art unfolds in the imagination of the viewers, be they collectors, critics, artists or children. Consilient Art invites reflection on stressful transformations occurring the world we live in today—contemplation being an antidote to global anxiety. Artists blaze new aesthetic pathways with Consilient Art offering creativity and contemplation as critical ingredients to the solutions needed to avert a global crisis—art as an agent of cultural resilience. With our world system on the brink of chaos—the toxic contamination of the biosphere and the collapse of global socio-economics—we are clearly facing a challenge to chose between and Age of Chaos or an Age of Transformation.

Appendix

Cover: Jackson Pollock, Lavender Mist © 2010 The Pollock-Krasner
Foundation / Artists Rights Society
(ARS), New York

Fig 3, Page 28. Barnett Newman, Onement 1 © 2010 The Barnett
Newman Foundation, New York / Artists Rights
Society (ARS), New York

Fig 6, Page 40, Marcel Duchamp, Mona Lisa © 2010 Artists Rights
Society (ARS), New York / ADAGP, Paris /
Succession Marcel Duchamp

Fig 8, Page 45, Damien Hirst, The Physical Impossibility Of Death in
the Mind of Someone Living; Fig 11, Page 49, Damien Hirst, For
the Love of God © 2010 Hirst Holdings Limited and Damien Hirst.
All rights reserved, ARS, New York / DACS, London

Fig 9, Page 46, Picasso, Bull; Fig 17, Page 60, Picasso, Self-Portrait ©
2010 Estate of Pablo Picasso / Artists Rights Society (ARS), New
York

Fig 22 Hanna Hoch, Cut with the Dada Kitchen Knife through the Last
Weimar Beer-belly Cultural Epoch in Germany © 2010 Artists
Rights Society (ARS), New York / VG Bild-Kunst, Bonn

Fig 25 Michael Tracy *Cruz de la paz Sagrada, 1980*, Menil Collection
Houston Photo credit to Hickey-Roberston courtesy of the Menil
Foundation

Bibliography

Bataille, Georges *The Prehistoric Painting*, Lascaux or the birth of art, Skira, 1955

Berger, J. *Ways of Seeing*. London: British Broadcasting Corporation, 1972.

Bowlby, J. *Attachment and Loss Series*. New York, NY: Basic Books, 1983.

Bowlby, J. *A Secure Base*. New York, NY: Basic Books, 1990.

Carroll, S. *Endless Forms Most Beautiful: The New Science of Evo Devo*. New York NY: W. W. Norton, 2006.

Dalai Lama, *The Four Noble Truths* New York, NY: Thorsons, 1998.

Danto, A. C. *The End of Art*. Princeton, NJ: Princeton University Press, 1998.

Darwin, C. *Origin of the Species*. New York, NY: Sterling, 2009.

Darwin, C. *The Expression of Emotions in Man and Animals*. New York, NY: Oxford University Press, 2009

Diamond, J. *Guns, Germs, and Steel*. New York, NY: W. W. Norton, 1999.

Donald, M. *Origins of the Modern Mind*. Cambridge, MA: Harvard University Press, 1993.

Donald, M. "Art and Cognitive Evolution." In Turner, M. *The Artful Mind*. New York, NY: Oxford University Press, 2006.

Frankl, V. *Man's Search for Meaning*.

Harmon-Jones, E., ed. *Social Neuroscience*. New York: Guilford, 2007.

Kuhl, P. *Scientist in the Crib*. New York, NY. Harper Collins, 1999.

Kuspit, D. *The End of Art*. New York: Cambridge University Press, 2005.

Lévi-Strauss, C. *Myth and Meaning*. New York: Shocken Books, 1979.

Sawyer, K. *Group Genius*. New York, NY: Basic Books, 2007.

Solso, R. L. *The Psychology of Art and the Evolution of the Conscious Brain.* Cambridge, MA: MIT Press, 2003.

Strick, J. *The Sublime Is Now: The Early Work of Barnett Newman.* New York: Pace Wildenstein, 1994.

Thompson, W. I. *Coming into Being.* New York, NY: St. Martin's Press, 1996.

Thrangu, K. *An Ocean of Ultimate Meaning.* Boston, MA: Shambala Publications, 2004.

Tsering, T. *The Four Noble Truths.* Somerville, MA: Wisdom Publications, 2005.

Valsiner, J. *Culture in Mind and Society.* Newbury Park, CA: Sage Publications, 2007.

Vygotsky, L. S. *Mind in Society.* Cambridge, MA: Harvard University Press, 1978.

Wallerstein, I. *World-Systems Analysis.* Durham, NC: Duke University Press, 2004.

Wilson, E. O. *Consilience: The Unity of Knowledge.* New York, NY: Vintage, 1999.

Winnicott, D. W. *Playing and Reality.* New York: Routledge, 2005.

Zaidel, D. *Neuropsychology of Art.* Hove and New York: Psychology Press, 2005.

www.ingramcontent.com/pod-product-compliance
Lightning Source LLC
Chambersburg PA
CBHW022108170526
45157CB00004B/1541